ENERGY SCIENCE, ENGINEERING, AND TECHNOLOGY

EXPLORING RENEWABLE AND ALTERNATIVE ENERGY USE IN INDIA

ENERGY SCIENCE, ENGINEERING, AND TECHNOLOGY

Additional books in this series can be found on Nova's website
under the Series tab.

Additional E-books in this series can be found on Nova's website
under the E-books tab.

ENERGY SCIENCE, ENGINEERING, AND TECHNOLOGY

EXPLORING RENEWABLE AND ALTERNATIVE ENERGY USE IN INDIA

JONATHAN R. MULDER
EDITOR

Nova Science Publishers, Inc.
New York

Copyright ©2012 by Nova Science Publishers, Inc.

All rights reserved. No part of this book may be reproduced, stored in a retrieval system or transmitted in any form or by any means: electronic, electrostatic, magnetic, tape, mechanical photocopying, recording or otherwise without the written permission of the Publisher.

For permission to use material from this book please contact us:
Telephone 631-231-7269; Fax 631-231-8175
Web Site: http://www.novapublishers.com

NOTICE TO THE READER

The Publisher has taken reasonable care in the preparation of this book, but makes no expressed or implied warranty of any kind and assumes no responsibility for any errors or omissions. No liability is assumed for incidental or consequential damages in connection with or arising out of information contained in this book. The Publisher shall not be liable for any special, consequential, or exemplary damages resulting, in whole or in part, from the readers' use of, or reliance upon, this material. Any parts of this book based on government reports are so indicated and copyright is claimed for those parts to the extent applicable to compilations of such works.

Independent verification should be sought for any data, advice or recommendations contained in this book. In addition, no responsibility is assumed by the publisher for any injury and/or damage to persons or property arising from any methods, products, instructions, ideas or otherwise contained in this publication.

This publication is designed to provide accurate and authoritative information with regard to the subject matter covered herein. It is sold with the clear understanding that the Publisher is not engaged in rendering legal or any other professional services. If legal or any other expert assistance is required, the services of a competent person should be sought. FROM A DECLARATION OF PARTICIPANTS JOINTLY ADOPTED BY A COMMITTEE OF THE AMERICAN BAR ASSOCIATION AND A COMMITTEE OF PUBLISHERS.

Additional color graphics may be available in the e-book version of this book.

LIBRARY OF CONGRESS CATALOGING-IN-PUBLICATION DATA

Exploring renewable and alternative energy use in India / editor, Jonathan R. Mulder.
    p. cm.
Includes bibliographical references and index.
ISBN 978-1-61209-680-3 (hardcover)
1. Renewable energy sources--India. I. Mulder, Jonathan R.
TJ807.9.I4E97 2011
333.79'4130954--dc22
                    2011001538

Published by Nova Science Publishers, Inc. ✛ New York

CONTENTS

Preface — vii

Chapter 1   Indian Renewable Energy Status Report:
Background Report for DIREC 2010 — 1
National Renewable Energy Laboratory

Chapter 2   Resource Evaluation and Site Selection
for Microalgae Production in India — 123
Anelia Milbrandt and Eric Jarvis

Chapter 3   National Policy on Biofuels — 205
Government of India

Chapter Sources — 215

Index — 217

PREFACE

India has great potential to accelerate the use of its endowed renewable resources to power its growing economy with a secure and affordable energy supply. The Government of India recognizes that development of local, renewable resources is critical to ensure that it is able to meet both its economic and environmental objectives, and it has promoted this development through policy action. This new book discusses the ways in which India has already supported the growth of renewable energy technologies, the impact this has had on utilization of various technologies and the enormous remaining potential.

Chapter 1- India has great potential to accelerate the use of its endowed renewable resources to power its growing economy with a secure and affordable energy supply. The Government of India recognizes that development of local, renewable resources is critical to ensure that it is able to meet both its economic and environmental objectives, and it has promoted this development through policy action.

Chapter 2- India's growing demand for petroleum-based fuels associated with its growing economy and population presents challenges for the country's energy security given that it imports most of its crude oil from unstable regions in the world. This and other considerations, such as opportunities for rural development and job creation, have led to a search for alternative, domestically produced fuel sources. Biofuels derived from algal oil show considerable promise as a potential major contributor to the displacement of petroleum-based fuels, given its many advantages including high per unit land area productivity compared to terrestrial oilseed crops, utilization of low-quality water sources and marginal lands, and the production of both biofuels and valuable co-products.

Chapter 3- India is one of the fastest growing economies in the world. The Development Objectives focus on economic growth, equity and human well being. Energy is a critical input for socio-economic development. The energy strategy of a country aims at efficiency and security and to provide access which being environment friendly and achievement of an optimum mix of primary resources for energy generation. Fossil fuels will continue to play a dominant role in the energy scenario in our country in the next few decades. However, conventional or fossil fuel resources are limited, non-renewable, polluting and, therefore, need to be used prudently. On the other hand, renewable energy resources are indigenous, non-polluting and virtually inexhaustible. India is endowed with abundant renewable energy resources. Therefore, their use should be encouraged in every possible way.

In: Exploring Renewable and Alternative Energy Use in India   ISBN: 978-1-61209-680-3
Editor: Jonathan R. Mulder   © 2012 Nova Science Publishers, Inc.

Chapter 1

INDIAN RENEWABLE ENERGY STATUS REPORT: BACKGROUND REPORT FOR DIREC 2010

National Renewable Energy Laboratory

ACKNOWLEDGMENTS

This report was produced in collaboration between National Renewable Energy Laboratory (NREL) in the United States, German Technical Cooperation (GTZ), Renewable Energy Policy Network for the 21st Century (REN21) Secretariat in France, and Integrated Research and Action for Development (IRADe) in India. REN21 Secretariat has coordinated the process. The co-authors are extremely grateful for the masterful editorial work done by Paul Gilman and Mary Lukkonen. The co-authors thank Joshua Bauer for developing the graphics and designing the cover for this report. Finally, the co-authors would like to thank those who provided their feedback on earlier drafts of this report, in particular Nina Negic of Bridge to India Pvt. Ltd and Ron Benioff of NREL. Their suggestions and insights resulted in substantial improvements to the draft published here. Financing was provided by the German Federal Ministry for the Environment, Nature Conservation, and Nuclear Safety (BMU) and U.S. Department of Energy and Department of State.

LIST OF ACRONYMS

ADB	Asian Development Bank
APCTT	Asia Pacific Centre for the Transfer of Technology
ARTI	Appropriate Rural Technology Institute
BEE	Bureau of Energy Efficiency
BGPG	Biogas Distributed/Grid Power Generation Program
BHEL	Bharat Heavy Electricals Ltd
BLY	Bachat Lamp Yojana
BPL	below poverty line

CDM	Clean Development Mechanism
CEA	Central Electricity Authority
CER	certified emission reduction credit
CERC	Central Electricity Regulatory Commission
CFA	Central Finance Assistance
CFL	compact fluorescent lamp
ckm	circuit kilometers
CSP	concentrated solar power
CSR	corporate social responsibility
C-WET	Centre for Wind Energy Technology
DANIDA	Danish International Development Agency
DDG	Decentralized Distributed Generation
DIREC	Delhi International Renewable Energy Conference
DNES	Department of Non-conventional Energy Sources
DNI	direct normal irradiance
EBP	ethanol blended petrol
EGTT	Expert Group on Technology Transfer
EMD	earnest money deposit
EPC	engineer-procure-construct
EU	European Union
GBI	Generation-based Incentive
GDP	gross domestic product
GEF	global environment facility
GHG	greenhouse gas
GHI	global horizontal irradiance
GW	gigawatt (one billion watts)
GWh	gigawatt-hour
GWEC	Global Wind Energy Council
ha	hectare
IDBI	Industrial Development Bank of India
IDFC	Infrastructure Development Finance Corporation
IEA	International Energy Agency
IEC	International Electrotechnical Commission
IFC	International Finance Corporation
IGCC	integrated gasification combined cycle
INR	Indian Rupee
IREC	International Renewable Energy Conference
IREDA	India Renewable Energy Development Agency
JNNSM	Jawaharlal Nehru National Solar Mission
KfW	Kreditanstalt für Wiederaufbau (German Development Bank)
kW	kilowatt
kWh	kilowatt-hour
LPG	liquefied petroleum gas
MENA	Middle East and North Africa
MFI	microfinance institutions
MNRE	Ministry of New and Renewable Energy

MoP	Ministry of Power
MoRD	Ministry of Rural Development
MPNG	Ministry of Petroleum and Natural Gas
MSW	municipal solid waste
MTOE	million tonnes of oil equivalent
MVA	mega volt amperes
MW	megawatt (one million watts)
MWh	megawatt-hour
NABARD	National Bank for Agriculture and Rural Development
NAPCC	National Action Plan on Climate Change
NASA	U.S. National Aeronautics and Space Administration
NBMMP	National Biogas and Manure Management Program
NEP	National Electricity Policy
NGO	non-governmental organization
NREL	National Renewable Energy Laboratory
NTPC	National Thermal Power Corporation
NVVN	NTPC Vidyut Vyapar Nigam Ltd
PFC	Power Finance Corporation
PPA	Power Purchase Agreement
PV	photovoltaic
R&D	research and development
R&M	renovation and modernization
RE	renewable energy
REC	Renewable Energy Certificates
REDB	Rural Electricity Distribution Backbone
REEEP	Renewable Energy and Energy Efficiency Partnership
REN21	Renewable Energy Policy Network for the 21st Century
RGGVY	Rajiv Gandhi Grameen Vidyutikaran Yojana
RPO	Renewable Purchase Obligation
RVE	Remote Village Electrification
SEB	State Electricity Board
SEC	Solar Energy Centre
SEFI	Sustainable Energy Finance Initiative
SERC	State Electricity Regulatory Commission
SIPS	special incentive package scheme
SME	small- and medium-sized enterprise
SSE	Surface Meteorology and Solar Energy
SWERA	Solar and Wind Energy Resource Assessment
SWH	solar water heater
TERI	The Energy and Resources Institute
TWh	terawatt-hour
UNDP	United Nations Development Programme
UNEP	United Nations Environment Programme
UNFCCC	United Nations Framework Convention on Climate Change
USAID	United States Agency for International Development
USD	U.S. dollar

| VESP | Village Energy Security Programme |
| WIREC | Washington International Renewable Energy Conference |

EXECUTIVE SUMMARY

India has great potential to accelerate the use of its endowed renewable resources to power its growing economy with a secure and affordable energy supply. The Government of India recognizes that development of local, renewable resources is critical to ensure that it is able to meet both its economic and environmental objectives, and it has promoted this development through policy action.

The Indian economy has experienced tremendous growth over the past several years. Energy, in all its forms, underpins both past and future growth. For the Indian economy to continue this trajectory, India needs to address its energy challenges, which cross all sectors and impact all citizens. Electricity—both in terms of quality and access—is a key challenge.

The quality of the current electricity supply is impeding India's economic growth. Issues such as voltage fluctuation, frequency variation, spikes, black-outs, brown-outs, and other disruptions impact industrial, commercial, and residential consumers. The addition of grid-tied renewable power can help address these issues. The gap between the demand of customers connected to the grid and the available electricity supply reported by the Central Electricity Authority for 2009–2010 was almost 84 TWh, which is 10% of the total requirement. The peak demand deficit was more than 15 GW, corresponding to a shortage of 12.7%. Closing this gap will be critical for India to achieve its growth targets, and renewable energy has the potential to improve energy security and reduce dependence on imported fuels and electricity while striving to meet those goals.

Much of India's population is not experiencing the benefits of economic growth. The Government of India sees the provision of electricity to all as critical to inclusive growth. It recognizes off-grid renewable energy as a practical, cost-effective alternative to an expansion of grid systems in remote areas of the country.

To be able to provide adequate electricity to its population, India needs to more than double its current installed capacity to over 300 GW by 2017. Also, India's demand for oil in 2015 is expected to be 41% higher than in 2007 and almost 150% higher in 2030—needed primarily to feed a growing transportation sector. The Indian government is aware of the size and importance of the challenges and that success will depend on structural changes in the industry and on new technologies and business models. Renewable energy is well positioned to play a critical role in addressing this growing energy demand for the following reasons:

- *India has the natural resources.* India has abundant, untapped renewable energy resources, including a large land mass that receives among the highest solar irradiation in the world, a long coastline and high wind velocities that provide ample opportunities for both land-based and offshore wind farms, significant annual production of biomass, and numerous rivers and waterways that have potential for hydropower.

- *Renewable energy provides a buffer against energy security concerns.* India's use of its indigenous renewable resources will reduce its dependence on imported, expensive fossil fuels.
- *Renewable energy offers a hedge against fossil fuel price hikes and volatility.* Increased competition for limited fossil resources is projected to push prices up, while increased deployment of renewable technologies pushes prices down in line with technology improvements and economies of scale. For example, oil prices in 2030 are projected to be 46% higher than in 2010 while the investment costs for photovoltaic (PV) systems are expected to decrease to less than half of their 2007 levels over the same time period.
- *Off-grid renewable power can meet demand in un-served rural areas.* As a distributed and scalable resource, renewable energy technologies are well suited to meet the need for power in remote areas that lack grid and road infrastructure.
- *Renewable energy can be supplied to both urban and rural poor.* Renewable energy technologies offer the possibility of providing electricity services to the energy poor while addressing India's greenhouse gas (GHG) concerns and goals.
- *Renewable energy can support attainment of India' climate change goals.* Through its National Action Plan on Climate Change (NAPCC) and through its recently announced carbon intensity goal, India has made a commitment to addressing its carbon emissions.
- *India aims to be a global leader in renewable energy.* India's intention to play a leadership role in the emergent global green economy is driving investment in renewable energy technologies. Recognizing the magnitude of the potential demand for renewable energy, India is attracting significant investment in renewable energy.

Renewable energy represents an area of tremendous opportunity for India, and this report discusses the ways in which India has already supported the growth of renewable energy technologies, the impact this has had on utilization of various technologies, and the enormous remaining potential. The report is meant for those who want to better understand the role renewable energy has had to date in India, the policies that have been implemented to support renewable energy deployment, and the potential for renewable energy technologies to expand their contribution to India's growth in a way that is consistent with India's developmental and environmental goals.

Section 1, Indian Energy and Climate Change Status, gives a brief overview of the climate change policies that influence energy use in India, gives an overview of India's energy sector, and discusses the specific conditions and challenges of the Indian power market.

India currently emits approximately 4% of global GHG emissions. However, its per capita emissions are only one-quarter of the global average and less than one-tenth of those of most developed nations. India has committed to reducing the emissions intensity of its economy to 20%–25% below 2005 levels by 2020 and has pledged that per capita GHG emissions will not exceed those of industrialized nations.

There are two primary climate-focused instruments that are influencing deployment of renewable energy technologies in India. The first is NAPCC, which was released in 2008 with the aim of promoting development goals while addressing GHG mitigation and climate change adaptation. NAPCC suggests that up to 15% of India's energy could come from renewable sources by 2020. The NAPCC includes eight focused missions, one of which is dedicated to solar energy (the others concern energy efficiency, water, sustainable habitat, and related topics). In addition, the Clean Development Mechanism (CDM) of the Kyoto Protocol is supporting development of renewable energy projects in India. As of September 2010, there were over 500 registered CDM projects in India, and these are dominated by renewable energy projects.

Chapter 2, The Status of Renewable Energy in India, describes the role of renewable energy in the overall energy sector in India and more specifically in industrial end use, transportation, and electricity generation; it presents the growth trends of renewable energy in India and the institutional and policy environment that is currently supporting continued deployment of renewable energy technologies.

The Government of India's Ministry of New and Renewable Energy (MNRE) reports that, as of June 2010, India has over 17.5 GW of installed renewable energy capacity, which is approximately 10% of India's total installed capacity. Wind represents 11.8 GW, small hydro represents 2.8 GW, and the majority of the remainder is from biomass installations. PV installations have reached 15 MW of cumulative capacity installation for both on- and off-grid applications. In the current, Eleventh Five-Year Plan (Eleventh Plan; 2007-20 12), the Government of India targets capacity additions of 15 GW of renewable energy, which, if achieved, would bring the cumulative installed capacity to over 25 GW in 2012.

In 1992, the Government of India established MNRE, the world's first ministry committed to renewable energy. MNRE is dedicated to expanding contributions of renewable energy in all of India's end-use sectors and undertakes policy and planning activities to that end. MNRE also supervises national-level renewable energy institutes such as the Solar Energy Centre and the Centre for Wind Energy Technology. The Indian Renewable Energy Development Agency (IREDA) provides financial support and innovative financing for renewable energy and energy efficiency projects with funds from the Indian government and multilateral lending agencies. IREDA also administers the central government's renewable energy incentive programs. Other government institutions with direct responsibilities that extend into renewable energy include several units under the Ministry of Power, the Planning Commission, and the Prime Minister's Council on Climate Change.

The Government of India has enacted several policies to support the expansion of renewable energy. Those that apply to more than one renewable technology include the following (technology-specific policies are discussed in the relevant chapters):

- Electricity Act 2003: Mandates that each State Electricity Regulatory Commission (SERC) establish minimum renewable power purchases; allows for the Central Electricity Regulatory Commission (CERC) to set a preferential tariff for electricity generated from renewable energy technologies; provides open access of the transmission and distribution system to licensed renewable power generators.
- National Electricity Policy 2005: Allows SERCs to establish preferential tariffs for electricity generated from renewable sources.

- National Tariff Policy 2006: Mandates that each SERC specify a renewable purchase obligation (RPO) with distribution companies in a time-bound manner with purchases to be made through a competitive bidding process.
- Rajiv Gandhi Grameen Vidyutikaran Yojana (RGGVY) 2005: Supports extension of electricity to all rural and below poverty line households through a 90% subsidy of capital equipment costs for renewable and non-renewable energy systems.
- Eleventh Plan 2007–2012: Establishes a target that 10% of power generating capacity shall be from renewable sources by 2012 (a goal that has already been reached); supports phasing out of investment-related subsidies in favor of performance-measured incentives.

As of April 2010, 18 states had established RPOs or had draft regulations under consideration with RPO requirements ranging from 1% to 15% of total electricity generation. In January 2010, CERC announced the terms of a tradable Renewable Energy Certificate (REC) program. Under this program, generators choose between selling the renewable electricity generated at a preferential tariff and selling the electricity generated separately from the environmental benefits. The environmental attributes can be exchanged in the form of RECs, which will be issued by a central agency set up for administration of this program. RPOs are not yet enforced.

Section 3, Wind Power, describes India's wind potential and development to date, policies supporting the growth of wind energy in India, and key opportunities as India's wind sector continues to expand.

MNRE's official estimate of India's on-shore wind capacity potential is over 48 GW. Industry associations assert that taking into account hub heights greater than 50 m and improving conversion efficiencies from technology advancements and aggressive policy action, the potential is much greater with ranges between 65 and 242 GW. India ranked fifth globally in total installed wind capacity in 2009, and as of March 2010, India had realized almost 12 GW of installed wind capacity. In addition to the sizable on-shore potential identified for India, the long coastline and prevailing wind patterns suggest substantial potential for offshore capacity, though no systematic assessment of offshore potential has yet been done.

The efficiency of India's existing wind plants is somewhat lower than in many of the other countries leading in wind. Technology and market factors combined with a shift from capacity-based incentives to generation-based incentives are expected to increase the efficiency of new wind plants.

Section 4, Solar Power, discusses the status of three solar technologies in India: solar PV, concentrating solar power (CSP), and solar water heating (SWH).

In June 2010, cumulative installed capacity of solar PV in India reached 15.2 MW, of which 12.3 MW was grid-tied (less than 0.1% of grid-tied renewable energy capacity in India) and 2.9 MW was off-grid (0.7% of off-grid renewable energy capacity in India). The recent Jawaharlal Nehru National Solar Mission (JNNSM) aims to dramatically increase installed PV through attractive feed-in tariffs and a clear application and administration process. Phase 1 of JNNSM targets additions of 500 MW of grid-tied and 200 MW of off-grid PV capacity by 2013. PV installations of 365 MW are approved for the state of Gujarat and 36 MW for the state of Rajasthan. By the end of Phase 3 in 2022, India plans to have 10,000 MW of grid- tied and 2,000 MW of off-grid PV.

JNNSM reaches beyond installed capacity to also target the growth of India's indigenous PV industry, which has historically supplied the export market. Annual Indian module production is expected to exceed 2,500 MW by 2015, and JNNSM targets an annual production of 4,000–5,000 MW by 2022.

There are currently no CSP plants in India. However, JNNSM targets the introduction and substantial additions of CSP. Estimates across India show large areas with sufficient radiation for the development of CSP plants, and one analysis foresees the technical potential for CSP generation in India at almost 11,000 TWh per year. JNNSM envisions 500 MW of CSP by 2013 and 10,000 MW by 2022. India is on track to meet the first goal—currently there are 351 MW of approved CSP projects in Gujarat and another 30 MW in Rajasthan.

SWH technology presents an opportunity to avoid using electricity or liquefied petroleum gas for heating water in India. It has applications in the residential, commercial, and industrial sectors. India has approximately 3.5 million m^2 of installed SWH collector area out of an estimated potential of 40 million m^2. Several programs are in place to accelerate deployment of SWH, including the provision of low-interest loans and incorporation of SWH into the building code.

Section 5, Small Hydro, presents the status of small hydropower in India, defined by the Government of India as hydropower plants up to 25 MW, including the identified potential and contribution to India's rural electrification efforts.

MNRE has estimated India's small hydro potential at more than 15,000 MW and is constantly revising this number upwards as new sites are identified; more than 40% of this potential has been identified in four northern, mountainous states. Capital incentives are in place to support installation of new small hydropower plants, as well as to renovate and modernize existing plants. In addition to direct subsidies at the state and national level, small hydro plants may also qualify for low-interest loans and income tax exemptions. CERC also offers preferential tariffs for small hydro plants with the tariff amount and time period varied depending on project size and location.

High capacity factors have been observed at small hydro plants in remote, mountainous regions of India, and small hydro has the lowest levelized costs of energy of any renewable technology in India. These factors have contributed to small hydro's importance in India's rural electrification efforts, with both federal and state programs supporting small hydro as a means of supplying electricity to villages where providing access to the central grid is challenging.

Section 6, Bioenergy, covers bioenergy in India and discusses the status and potential for biogas, solid biomass, and biofuels.

Traditional biomass, such as wood and cow dung, have historically played an important part in India's energy supply, and they still supply cooking energy to almost all of India's rural population. By utilizing organic waste and agricultural output, India can incorporate modern bioenergy as a substantial part into its future energy mix.

There are approximately 4 million installed household biogas plants in India with almost 4,000 additional units supplying household clusters or villages; cattle manure is the primary feedstock for these household plants. MNRE estimates that available cattle manure could support approximately 12 million household biogas plants. Larger-scale biogas facilities use industrial wastewater to generate electricity, and the 48 such plants in India have an aggregate installed capacity of 70 MW. MNRE estimates the total capacity potential for industrial biogas to be 1.3 GW. National and state programs support expansion of biogas through technical capacity building, financial incentives, and demonstration projects, with additional deployment supported through the CDM.

Solid biomass is used in India either in direct combustion or gasification to generate power or for cogeneration of power and heat. MNRE estimates that surplus biomass could support 25 MW of installed electricity-generating capacity and that cogeneration capabilities added to existing industries requiring process heat could add 15 GW more electricity-generating capacity to the grid. Policies in place to support biomass power generation include capital and interest subsidies and tax exemptions, and CERC and several states offer a preferential tariff for electricity generated from biomass power or cogeneration plants.

Liquid biofuels, ethanol and biodiesel, are used to substitute petroleum-derived transportation fuels. Ethanol in India is largely produced by the fermentation of molasses, a by-product of the sugar industry, and from biodiesel produced from non-edible oilseeds. India is conducting research in the area of cellulosic biomass conversion for ethanol production. Making use of one-third of the 189 million tonnes of surplus biomass could yield approximately 19 billion liters of gasoline equivalent each year, the equivalent of India's entire gasoline consumption. Support for liquid biofuels comes in the form of subsidized loans for ethanol production facilities, tax exemptions for biodiesel, financial incentives promoting the cultivation of Jatropha and other non-edible oil seeds, and research and development programs.

Section 7, Decentralized Energy, presents the role of renewable energy in India's decentralized energy systems including rural electrification and captive power for industry.

According to the International Energy Agency, in 2008, more than 400 million Indians did not have access to electricity, with electrification rates of 93.1% and 52.5% in urban and rural areas, respectively. There are several programs at the national level to promote electrification of remote villages and below poverty line households. Renewable technologies deployed to this end include family biogas plants, solar street lights, solar lanterns, solar PV systems, and micro-hydro plants. The Indian government has initiated several programs, policies, and acts that focus on the development of rural energy, economy, and electrification to improve rural livelihood with the help of renewable energy. Successful business models have been designed to address shortcomings of global electrification efforts to increase sustainability and user satisfaction. Among these is a model for village electrification using biomass gasifiers where an energy service company, entrepreneur, or other actor provides system maintenance and operations and collects feedstock for use from households and businesses. Finding viable business solutions to supplying India's non-electrified households is perhaps the greatest single challenge and opportunity of the renewable energy market.

In addition to using renewable energy to provide electricity to poor and remote citizens, renewable technologies are being used by industries that generate electricity at their facilities to supplement grid power and to use when power from the grid is unavailable. At the end of March 2008, the total renewable installed capacity for captive power plants greater than 1 MW was 305 MW, almost all of which was wind capacity. In addition to wind, solar thermal plants show good potential for captive power supply by integrating with the steam turbines already in place at cogeneration facilities. Industries can also deploy solar PV systems and use biodiesel in onsite generators to help power their operations.

Section 8, Technology Transfer, discusses the relevance of international renewable energy technology to India as well as roles of the public and private sectors.

Technology transfer is considered an important element of India's low-carbon growth strategy. Though India is a strong international competitor in areas like wind and PV, a domestic base for other renewable technologies is lacking, and much of the research and innovation is happening outside the country. India presents opportunities across the value chain including research and development, component manufacturing, investment, project development, power production, service and maintenance, and training and education. The private sector requires a stable investment environment and cost competitiveness to capitalize on these opportunities. Actions of the Indian government and the international community can contribute to the conditions necessary for accelerated entry of international actors into India's renewable energy sector.

Section 9, Financing Renewable Energy in India, presents the current renewable energy investment climate in India.

In 2009, India ranked eighth globally for clean energy investments and is ranked as the fourth most attractive country for renewable energy investment, only behind the United States, China, and Germany. The majority of renewable energy financing in India has been asset finance (for renewable energy generation projects), accounting for 70% in 2009 after a drop from levels in previous years. The majority of this went to the wind sector, though with the release of JNNSM and the National Biofuels Policy, greater shares may go to the solar and bioenergy sectors in the future. International and bilateral finance institutions are involved in clean energy financing in India, providing private sector financing, technology financing, and more specialized support. Renewable energy in India is also receiving financial support through the CDM. India accounts for more than 20% of registered projects under CDM, and the majority of these are renewable energy projects. The Government of India and nongovernmental organizations have historically been the largest funders of clean energy in India, and they still are active in their support. IREDA administers the federal government's revolving renewable energy fund and administers much of the international funding as well. The balance of renewable energy finance in India is made up of various sources including venture capital and commercial financing.

Section 10, Enabling Environment, concludes the paper with a discussion of factors for continued renewable energy growth in India and elsewhere and the role of the International Renewable Energy Conference series.

The International Renewable Energy Conference process seeks to bring together renewable energy stakeholders from across the world to share best practices for renewable energy policy, financing, and human capacity development. The Delhi International Renewable Energy Conference (DIREC) will build on the progress of the previous three in this series of ministerial-level conferences and will focus on scaling up and mainstreaming renewable energy in the context of energy security, climate change, and economic development, all of which are important themes to the Government of India. The conference will include discussions on key factors that are critical to widespread deployment of renewable energy technologies: the policy landscape needed to affect a substantial scale-up of renewable energy globally, the innovative financing that may also facilitate accelerated renewable development, and technology development and cost trends that are expected in the coming years, some of which are also discussed in this chapter.

India seeks to become an international leader in renewable energy to support its goals of energy access for the sectors of the population currently un-served; fuel its economy, which is returning to an annual growth of nearly 9% as experienced before the recent downturn; enhance energy security against price and supply disruptions; and demonstrate leadership on the global stage in mitigating GHG emissions. JNNSM is India's most ambitious renewable energy initiative. If its goal of achieving grid parity of solar power by 2022 can be achieved, this energy source can quickly become one of the pillars of India's energy mix. India has substantial on-shore wind generating capacity, with additional potential for development. Preliminary investigation of India's off-shore wind speeds suggests off-shore wind also has strong potential for contributing to India's power generation mix. Small hydro remains a cost-effective means for providing electricity in many areas of India, including the remote Himalayas where connection to the central grid is impractical. The potential for biopower generation from waste materials is far from being exhausted, and these plants could add much-needed capacity to the grid. Indigenous biofuel production, including from emerging technologies such as cellulosic ethanol and algal biodiesel, could aid India's growth without developing an overreliance on imported oil. Maximum utilization of India's resources will require several elements acting in concert: a supportive policy environment, ample financing opportunities, and technology advancement. India has demonstrated commitment and leadership in tacking the challenges and is poised to accelerate this transition to a robust renewable energy future.

1. INDIAN ENERGY AND CLIMATE CHANGE STATUS

Section Overview

In 2008, India accounted for 17.7% of the world population but was the fifth-largest consumer of energy, accounting for 3.8% of global consumption. India's total commercial energy supply is dominated by coal and largely-imported oil with renewable energy resources contributing less than 1% (this does not include hydro > 25 MW). Coal also dominates the power generation mix, though renewable resources now account for approximately 10% of installed capacity. The current power-generating capacity is insufficient to meet current demand, and in 2009–2010, India experienced a generation deficit of approximately 10% (84 TWh) and a corresponding peak load deficit of 12.7% (over 15 GW). India's frequent electricity shortages are estimated to have cost the Indian economy 6% of gross domestic product (GDP) in financial year 2007–2008. To power the economic growth currently being targeted, it is estimated that India will need to more than double its installed generating capacity to over 300 GW by 2017. In recent years, control over generating facilities has shifted from being dominantly controlled by the states to the federal government and private entities, including those who have set up captive power plants to power their industrial facilities. The private sector is dominant in renewable energy generation.

India's energy future will not just be shaped by the central grid and large-scale generating facilities fueling industrial growth but also by the goal of increasing the well-being of India's poor populations by providing electricity access to the approximately 400 million citizens without. The Government of India recognizes that development of local, renewable resources is critical to ensure that India is able to meet social, economic, and environmental objectives and has supported the development of renewable energy through several policy actions.

Energy planning in India is taking place in the context of climate change negotiations. India participates in the international climate negotiation process, has pledged to reduce its economy's greenhouse gas (GHG) intensity, and has pledged that its per capita emissions will not exceed those of developed nations. India has implemented a National Action Plan on Climate Change (NAPCC), which suggested that 15% of energy could come from renewable sources by 2020. The NAPCC has eight National Missions, one of which is focused specifically on renewable energy: The Jawaharlal Nehru National Solar Mission (JNNSM). India is an active participant of the Clean Development Mechanism (CDM) with the second largest number of projects registered among all countries participating, the majority of which are renewable energy projects.

1.1. Climate Change and Existing Relevant Policies

While India's first priority is to ensure a sufficient and stable supply of power by expanding fossil-fuel-based generation and installing new large hydro and nuclear power plants, the following considerations contribute to the case for developing renewable energy projects:

- The use of India's indigenous renewable resources will reduce India's dependence on imports of fossil fuels.
- Increased competition for limited fossil resources is projected to push prices up, while increased deployment of renewable technologies will push prices down in line with technology improvements and economies of scale. For example, oil prices in 2030 are projected to be 46% higher than in 2010[1] while the investment costs for photovoltaic (PV) systems are expected to decrease to less than half of their 2007 levels over the same time period.[2]
- Renewable energy technologies are well-suited to meet India's need for power in remote areas that lack grid and road infrastructure due to the distributed nature of resources and the scalability of system design.
- India's intention to play a leadership role in the emergent global renewable energy economy is driving investment in renewable energy technologies.

- Renewable energy technologies offer the possibility of providing electricity services to the energy poor while addressing India's GHG concerns and goals.

While no official goal for stabilizing GHG emissions has been set, India has declared that per capita GHG emissions will not be permitted to exceed those of industrialized countries[3] and has committed to reducing the emission intensity of its GDP by 20%–25% by 2020 from the 2005 level.[4] The main ways of achieving this reduction are improving fossil fuel power generation through new technologies such as supercritical and ultra-supercritical coal plants and integrated gasification combined cycle (IGCC), reducing aggregate technical and commercial losses, improving public transport, and increasing end-user efficiency. Given the shortage and cost of fossil resources, these measures also make sense from a purely economic perspective. Reducing emissions beyond the capability of these measures will require India to shift to low-carbon technologies that have a much higher initial capital commitment and much lower rate of return.[5]

Currently, India's emissions are around 1 ton of CO_2 per person per year. The global per capita average is 4.2 tons with most industrialized countries emitting 10–20 tons per person per year. Nevertheless, because of its large population, India already contributes around 4% to global emissions.[6]

1.1.1. National Action Plan on Climate Change

In June 2008, India released the NAPCC to promote development goals while addressing climate change mitigation and adaptation.[7]

One of the eight National Missions outlined in the NAPCC, JNNSM (described in Chapter 4), specifically focuses on solar energy and its role to minimize future emissions while expanding development opportunities throughout the country.

The other seven National Missions are the National Mission for Enhanced Energy Efficiency, the National Mission on Sustainable Habitat, the National Water Mission, the National Mission for Sustaining the Himalayan Ecosystem, the National Mission for a Green India, the National Mission for Sustainable Agriculture, and the National Mission on Strategic Knowledge for Climate Change.

The NAPCC suggested that as much as 15% of India's energy could come from renewable sources by 2020.

1.1.2. Clean Development Mechanism

Under the Kyoto Protocol, the CDM allows for projects in developing countries that result in a reduction of GHG emissions to earn certified emission reduction credits (CERs).[8] CERs are carbon credits issued under the CDM that each represent 1 ton of carbon dioxide equivalent. CERs can be traded or sold and eventually used by industrialized countries that have ratified the Kyoto Protocol to meet parts of their emissions reduction targets. To qualify to receive CERs, the project must first be approved by the Designated National Authorities of the project host country, and then a public registration and issuance process with the UNFCCC must be completed. The additional funds generated through the sale or trade of CERs can improve the financial viability of clean energy projects.

Although there has been some criticism of the CDM process (e.g., the additional financial and technical criteria necessary or the long and costly project certification process), India has been at the forefront of receiving CDM benefits. As of January 2010, 1,551 Indian projects have been granted host country approvals. The projects are in the fields of energy efficiency, fuel switching, industrial processes, municipal solid waste (MSW), and renewable energy. Pending registration by the CDM executive board, India will generate a total of 627 million CERs before the end of 2012. At a price of INR 500 (USD 10) per CER, they would be worth INR 313.5 billion (USD 6.27 billion). In 2009, 478 of the world total 2,011 registered projects were generated by India. Only China had registered more.[9] As of September 2010, India had 532 registered CDM projects, 426 of which were in the energy sector, with the overwhelming majority comprised of renewable energy projects.[10]

1.2. Energy Sector Overview

The electricity intensity of the Indian economy—the percentage growth of electricity consumption that correlates with 1% of economic growth—fell from approximately 3.14% in the 1950s to 0.97% in the 1990s.[11] In 2007, it was at 0.73%. The main reason for this reduction is that India's growth until now was based more on the service sector (with an electricity intensity of only 0.11%) than on growth in industrial production (with an electricity intensity of 1.91%).[12] Today, for each 1% of economic growth, India needs around 0.75% of additional energy.[13] The Planning Commission of India, which coordinates Indian long-term policy, analyzes different scenarios; one scenario assessed that this value could fall to 0.67% between 2021–2022 and 2031–2032.[14]

India is facing a formidable challenge to build up its energy infrastructure fast enough to keep pace with economic and social changes. Energy requirements have risen sharply in recent years, and this trend is likely to continue in the foreseeable future. It is driven by India's strong economic and population growth as well as by changing lifestyle patterns. Growth and modernization essentially follow the energy-intensive Western model of the 19th and 20th centuries, in which economic growth correlates with a comparable growth in the energy use.

For GDP annual growth of 8%, the Planning Commission estimates that the commercial energy supply would have to increase at the very least by three to four times by 2031–2032 and the electricity generation capacity by five to six times over 2003–2004 levels.[15] In 2031–2032, India will require approximately 1,500–2,300 million tonnes of oil equivalent (MTOE) to cover its total commercial energy needs.[16] The Indian government by itself does not have sufficient financial resources to solve the problem of energy shortages. It must rely on cooperation with the private sector to meet future energy requirements. This opens up interesting market opportunities for international companies.

In the study "Powering India: The Road to 2017," the management consultancy McKinsey calculated that goals specified by the Indian government have been set too low.[17] The study argues that the peak demand will be 315 and 335 GW by 2017 instead of by only 213 GW as estimated in 2007 by the Central Electricity Authority (CEA) and 226 GW (assuming 8% annual GDP growth) estimated in the Integrated Energy Policy (see Figure 1-1). Accordingly, by 2017, India would require a total installed capacity of 415–440 GW in order to be able to reliably meet this demand. This means that over the next 9 years, the country would have to install more than twice as much capacity as it has been able to install over the last 60 years (159 GW). In order to achieve this, the speed with which the new capacities are built must increase five-fold. Such a massive increase requires a fundamental restructuring of the entire power market, which the government has begun through enactment of the Electricity Act (discussed in Section 1.3 and Chapter 2).

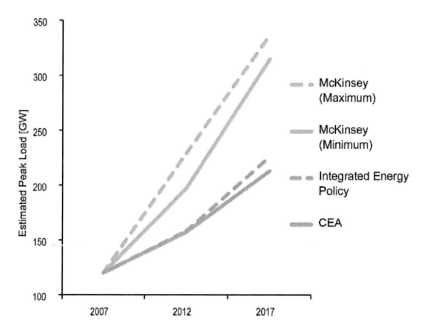

Figure 1-1. Scenarios for the development of necessary peak load capacities[18].

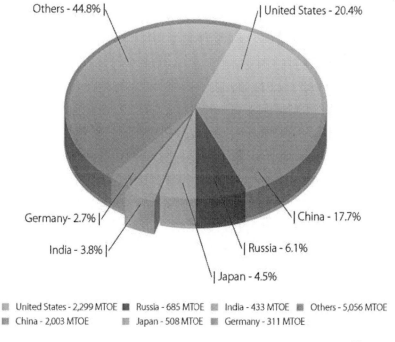

Figure 1-2. Worldwide consumption of primary sources of energy by country (2008)[20].

1.2.1. Commercial Energy Consumption

As shown in Figure 1-2, India's share of the global commercial energy[19] consumption in 2008 was 3.8% (433 of 11,295 MTOE), increased from 2.9% over the past 10 years, thus making it the fifth largest consumer of commercial energy. By comparison, China holds 19.6% of the population and consumes 17.7% of commercial energy.

India's total consumption of commercial energy increased from 295 MTOE in the year 2000 to 433 MTOE in 2008 with an average annual growth rate of 4.9% (see Figure 1-3).

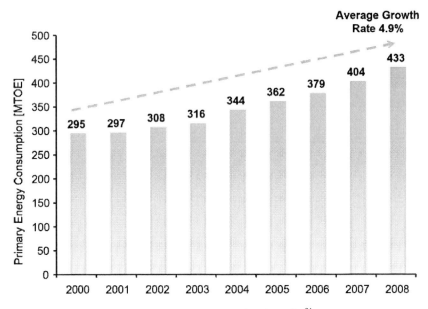

Figure 1-3. Development of commercial energy consumption in India [21].

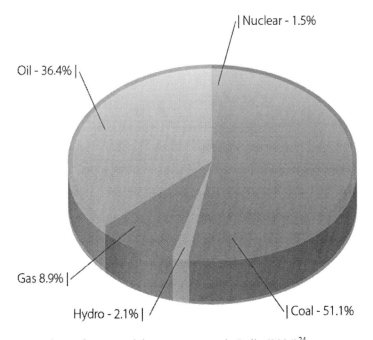

Figure 1-4. Percentage share of commercial energy sources in India (2004)[24].

Coal is by far the most important energy source for India; it provides more than half of the commercial energy supply. Oil, mostly imported, is the second most important source of energy, followed by gas and hydropower (see Figure 1-4). So far, nuclear (atomic) power covers only a small portion of the commercial energy requirement (approximately 1.5%). With less than 1%, renewable energy plays a minor role (this does not include hydro > 25 MW), and therefore, it is not even visible in Figure 1-3, though its share is projected to increase significantly. The traditional use of biomass (e.g., for cooking) has not been included here as a source of energy. However, the 2001 Census points out that approximately 139 million of the total 194 million households[22] in India (72%) are using traditional forms of energy such as firewood, crop residue, wood chips, and cow dung cakes for cooking.[23] The majority of these households are in rural areas. Firewood, used by approximately 101 million households, is the main cooking fuel in India.

1.3. The Power Market in India and the Role of Renewable Energy

While India has been making progress in different infrastructural areas such as the construction of roads and expansion of the telecommunication system, the power infrastructure has not kept pace with the growing requirements.

India's power market is confronted with major challenges regarding the quantity as well as the quality of the electricity supply. The base-load capacity will probably need to exceed 400 GW by 2017. In order to match this requirement, India must more than double its total installed capacity, which as of March 2010 was 159 GW.[25] Moreover, India's power sector must ensure a stable supply of fuels from indigenous and imported energy sources, provide power to millions of new customers, and provide cheap power for development purposes, all while reducing emissions. On the quality side, the electricity grid shows high voltage fluctuations and power outages in almost all parts of the country on many days for several hours.[26] According to the "Global Competitiveness Report," in 2009–2010 (weighted average), India ranked 110 among 139 countries in the category "Quality of Electricity Supply."[27]

The power deficit reported for 2008–2009 was almost 84 TWh, which is almost 10% of the total requirement; the peak demand deficit was more than 12.7% at over 15 GW.[28] The electricity undersupply in India is estimated to cost the economy as much as INR 34 (USD 0.68) to INR 112 (USD 2.24) for each missing kilowatt-hour. Thus, the total cost of the power deficit of 85 billion kWh in financial year 2007–2008 amounted to at least INR 2,890 billion (USD 58 billion), or almost 6% of the GDP.[29] Another report states that there is an approximately 7% decrease in the turnovers of Indian companies due to power cuts.[30] As a consequence, many factories, businesses, and private customers have set up their own power generation capacities in the form of captive power plants or diesel generators in order to ensure their power supply. This provides an attractive opportunity for renewable energy solutions; they compete not with power produced relatively cheaply by large coal plants but with much more expensive diesel back-up generators.

Until 1991, the Indian government monopolized the power market. There were only a few private actors, and the CEA had sole responsibility for giving techno-economic clearance to new plants. However, the public sector has been unable to cater to the growing demand for power, and in the future, investment requirements in the public sector will far exceed the resources. Current energy policies therefore place an emphasis on the integration of the private sector along the entire value chain: from the generation of power to transmission and distribution.

The Electricity Act 2003 displaced former energy laws and expanded them comprehensively.[31] The aim of the act was the modernization and liberalization of the energy sector through the implementation of a market model with different buyers and sellers. The main points included making it easier to construct decentralized power plants, especially in rural areas and for captive use by communities, and giving power producers free access to the distribution grid to enable wheeling. Producers could also choose to sell power directly to consumers rather than through the financially weak State Electricity Boards (SEBs). Through the Electricity Act, the different legal frameworks are to be unified at a state level to promote foreign direct investment in the country.

Given the long-term energy deficit and the growth trajectory of the Indian economy, the Indian investment community has responded positively. However, international investors are still hesitant. The largest barrier to more foreign private investment in the energy market is the energy price itself. In many customer sections and regions, they are too low to generate stable and attractive returns. Despite being an impractical drain on resources, the government has so far failed to adjust prices. The key reason is that cheap or free electricity is an important political token in a country where the majority of the population still lives on a very low income.[32]

1.3.1. Power Consumption

India's average power consumption per person was 733 kWh in 2009, and the average annual rate of increase since 2003 was 4.4%,[33] as shown in Figure 1-5.

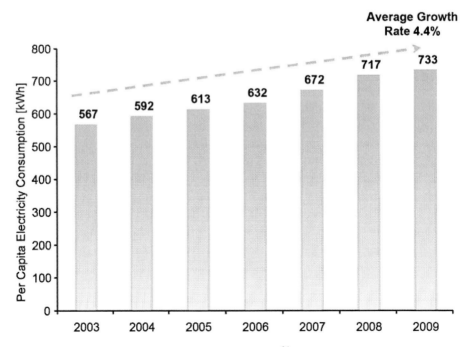

Figure 1-5. Per capita annual electricity consumption in India[34].

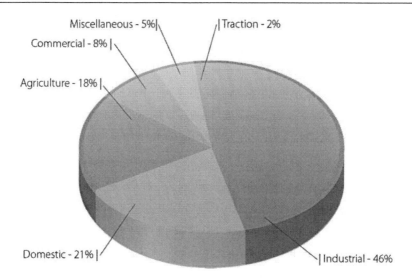

Figure 1-6. India electricity consumption sector-wise (utilities & non-utilities, 2008–2009)[36].

In 2008, a total of 596,943 GWh were consumed in India. The largest consumer was industry with 274,531 GWh (46%), followed by households with 124,562 GWh (21%), and agriculture with 107,835 GWh (18%). In the commercial sector (e.g., offices and shops), 48,047 GWh (8%) were consumed, 11,615 GWh (2%) in rail traffic, and 30,353 GWh (5%) in various other sectors.[35]

Between 1980 and 2009, energy consumption increased by almost seven times from 85,334 GWh to 596,943 GWh, which corresponds to an average annual growth rate of approximately 7.1%. The strongest increase was the consumption by private households, which increased by almost 14 times since 1980 at an average annual growth rate of 10%. The reason for this increase was the inclusion of several million new households, corresponding to the increase in electrical household appliances such as refrigerators and air conditioners. The agricultural share increased seven-fold at an annual growth rate of 7.6% between 1980 and 2008. The reason for a strong growth in the agricultural sector is, first, the inclusion of more rural areas, and second, the provision of power to farmers at reduced, or even free, rates in many areas. The consequence of this latter practice was the widespread purchase of cheap and inefficient water pumps that continue to run almost uninterrupted. The slowest growth in power consumption was seen in the industrial sector at 5.9% per year, which still corresponds to a five-fold increase.[37]

The main drivers for the strong growth in the demand for power are the overall economic growth, the power-intensive manufacturing industry that is growing disproportionately fast, the rapidly rising consumption in households due to the affordability of new electrical appliances, the planned provision of power to 96,000 currently un-electrified villages, and the provision of power for latent demand, which is currently unfulfilled because of frequent power cuts.[38]

1.3.2. Power Generation Capacity

The total power generation capacity in India in March 2010 was 159 GW. Of this, 64.3% was fossil-fuel-fired power plants (coal, gas, and diesel), 23.1% hydropower, 2.9% nuclear power, and 9.7% renewable energy.[39]

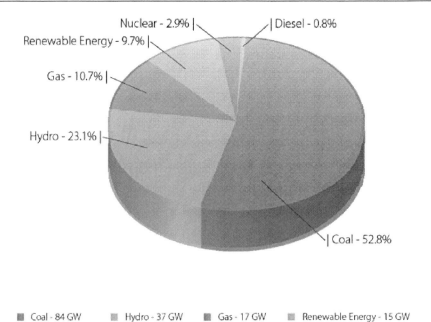

[Renewable energy includes small hydropower plants (< 25 MW), biomass gasification, biomass energy, urban and industrial waste energy, solar energy, and wind energy.].

Figure 1-7. Installed capacities for power generation in India according to energy source (March 2010)[40].

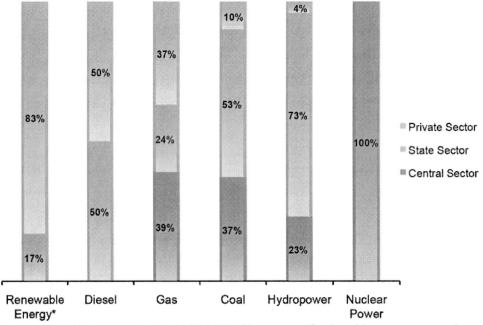

* Includes small hydropower plants (< 25 MW), biomass gasification, biomass energy, urban and industrial waste energy, solar energy, and wind energy.

Figure 1-8. Percentage of public and private sector power generation capacity by energy source (March 2010)[42].

The composition of the power sector has changed significantly in the last 30 years. The power generation capacity controlled directly by the central government has increased from 12% to 32%. At the same time, the fraction of generation capacity controlled by the individual states fell from 83% to 50%. Generation capacity controlled by the private sector more than tripled from 5% to 18%.[41] The private sector dominates in power generation from renewable energy sources.

The National Electricity Policy (NEP) assumes that the per capita electricity consumption will increase to 1,000 kWh by 2012. To cover this demand, the government is planning to add 78,700 MW of capacity during the Eleventh Five-Year Plan[43] (Eleventh Plan) ending March 2012. As of April 2010, 22,552 MW of new installation toward that goal had been achieved. There are further projects under construction with a total capacity of 39,822 MW. As per the mid-term plan review, capacity additions of 62,374 MW are likely to be achieved with a high degree of certainty and another 12,000 MW with best efforts.[44] Figure 1-9 shows India's capacity growth from the end of the Eighth Plan in 1997 to projections through the end of the Eleventh Plan.

Figure 1-10 shows the technology breakdown of the 78,700 MW targeted in the Eleventh Plan. The largest share of 59,693 MW is to be provided by thermal power plants. Additionally, 15,627 MW is to be provided by hydro and 3,380 MW by nuclear power. The central government undertakings, such as those of the National Thermal Power Corporation or the National Hydro Power Corporation, will contribute the most.[46]

In March 2009, the gross electricity generation[48] by utilities in India was 746.6 TWh. In addition, 95.9 TWh was generated by non-utilities and another 5.9 TWh were net imports.

The total generation available was thus 848.4 TWh, which corresponds to a rise of 3.3% as compared to the previous year.[49] As these figures show, the trend in growth rates is inadequate in view of the rapid increase in demand for power.

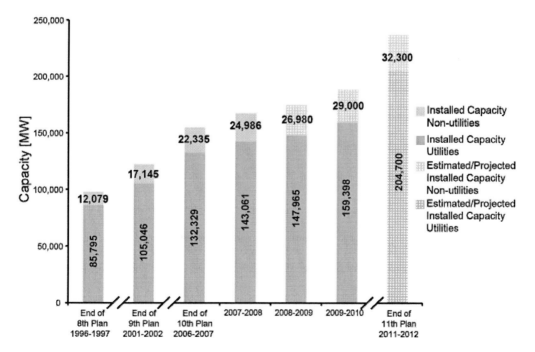

Figure 1-9. Development of installed electrical capacities of utilities and non-utilities in India [45].

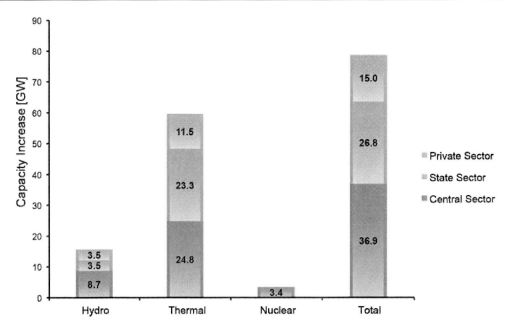

Figure 1-10. Forecast growth in capacity by the end of the Eleventh Plan according to sector (2012)[47].

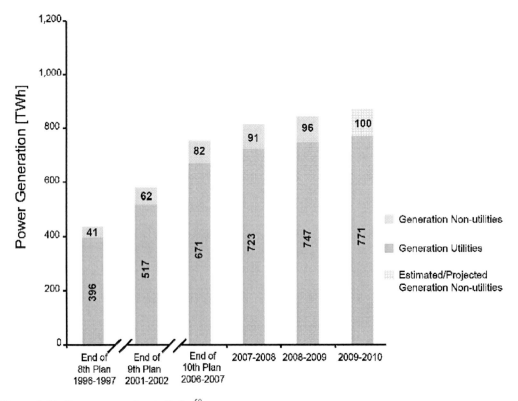

Figure 1-11. Power generation in India [50].

1.3.2.1. Electricity Generation Efficiency

Conventional thermal power generation in India faces three main challenges:

1. The low average conversion efficiency of the plants (30%).
2. The low quality of the coal itself, which has high ash content and a low calorific value (3,500–4,000 kcal/kg).[51]
3. The fixed electricity off-take price, which does not reward efficiency gains.

It is estimated that at least 25%–30% of the capacity in power plants in India is old and inefficient and operates at high heat rates and low utilization levels.[52]

To overcome these challenges, the Indian government has implemented a comprehensive program that includes a large-scale renovation and modernization (R&M) program for existing power plants, the promotion of supercritical technology for Ultra Mega Power Projects at pithead locations, the promotion of use of imported higher quality coal (from South Africa, Australia, and Indonesia) for coastal locations, the set-up of coal washing facilities for domestic coal, and the promotion of an IGCC technology for gas plants. Also, new power plant projects are being awarded via a competitive bidding process based on the lowest price offer for electricity sold to the grid.

Since 1985, nearly 400 units (over 40 GW) have been serviced through the R&M program. According to The Energy and Resources Institute (TERI), R&M could improve electricity generation by 30%, reduce emissions by 47%, and increase energy conversion efficiency by 23%.[53]

The R&M program currently faces two challenges to successful completion. First, the rising electricity demand makes it difficult to take plants off the grid for maintenance work. Second, sometimes the costs to repair or upgrade old power generation equipment exceed 50% of the costs of an entirely new plant. In such cases, repair is not economically viable. However, given the rising demand, such plants cannot be taken off the grid either. Although many newer, privately operated plants are more efficient than state-owned plants, there is still a technology deficit across the power generation sector, mainly with respect to the latest supercritical technology. The performance of India's existing supercritical power plants has so far failed to meet expectations.[54] This presents a great opportunity for international technical cooperation.

1.3.2.2. Transmission and Distribution Technical and Commercial Losses

India's electricity losses are among the highest in the world. In 2000, they were at a national average of 40%; in 2005, they were still at 30%; and in 2009, they were at 27%. The Indian government aims to bring losses down to 20% by 2012,[55] which is still far above the international average where figures well below 10% are considered appropriate.

Reasons for this high percentage of loss include a lack of investment into the grid, widespread power theft, and a lack of consumption monitoring. In addition, a large amount of electricity is freely provided to certain consumers (e.g., farmers). Both the losses and the wasteful distribution of free electricity weaken the financial position of the SEBs, thereby reducing their capabilities to invest in grid infrastructure.

In 2008, the government set up an INR 19.46 billion (USD 389 million) fund under the Restructured Accelerated Power Development and Reform Program from which SEBs could take out loans in order to finance grid improvements. The program was only moderately successful since the conditions of the loans did not create sufficient incentives.[56] There are, however, a few examples of successful reductions in electricity losses. Gujarat has managed to bring down losses from 30% in 2005 to 20% in 2009 by isolating the wasteful rural electricity consumption and then monitoring and optimizing it.[57] A second approach, which has so far been implemented only in large cities, is to involve the private sector by selling or franchising the existing infrastructure to private operators. This has been successful in Delhi, Mumbai, and Calcutta, all cities that have transmission losses that are lower than the national average.

Given that a large part of the country is not yet connected to the grid, India has the opportunity to create a more cost-effective and possibly climate-friendly decentralized electricity supply. India can even leapfrog ahead of developed countries by implementing renewable energy technologies where they are economically viable. The challenge is to develop a policy framework and functioning business models that attract investment into off-grid electricity generation.

2. THE STATUS OF RENEWABLE ENERGY IN INDIA

Section Overview

India has over 17 GW of installed renewable power generating capacity. Installed wind capacity is the largest share at over 12 GW, followed by small hydro at 2.8 GW. The remainder is dominated by bioenergy, with solar contributing only 15 MW. The Eleventh Plan calls for grid-connected renewable energy to exceed 25 GW by 2012. JNNSM targets total capacity of 20 GW grid-connected solar power by 2022. Renewable energy technologies are being deployed at industrial facilities to provide supplemental power from the grid, and over 70% of wind installations are used for this purpose. Biofuels have not yet reached a significant scale in India.

India's Ministry of New and Renewable Energy (MNRE) supports the further deployment of renewable technologies through policy actions, capacity building, and oversight of their wind and solar research institutes. The Indian Renewable Energy Development Agency (IREDA) provides financial assistance for renewable projects with funding from the Indian government and international organizations; they are also responsible for implementing many of the Indian government's renewable energy incentive policies. There are several additional Indian government bodies with initiatives that extends into renewable energy, and there have been several major policy actions in the last decade that have increased the viability of increased deployment of renewable technologies in India, ranging from electricity sector reform to rural electrification initiatives. Several incentive schemes are available for the various renewable technologies, and these range from investment-oriented depreciation benefits to generation- oriented preferential tariffs. Many states are now establishing Renewable Purchase Obligations (RPOs), which has stimulated development of a tradable Renewable Energy Certificate (REC) program.

2.1. Renewable Energy Share of Electricity

As of June 2010, India was one of the world leaders in installed renewable energy capacity, with a total capacity of 17,594 MW (utility and non-utility),[58] which represents approximately 10% of India's total installed electric generating capacity.[59] Of that total,

17,174 MW were grid-connected projects, and the remaining 2.4% of installed renewable capacity consisted of off-grid systems.[60]

The wind industry has achieved the greatest success in India with an installed capacity of 12,009 MW at the end of June 2010. India has also installed 2,767 MW of small hydro plants (with sizes of less than 25 MW each), 1,412 MW of grid-connected cogeneration from bagasse, and 901 MW of biomass-based power from agro residues. Waste-to-energy projects have an installed capacity of 72 MW.

India has off-grid renewable power capacities of 238 MW from biomass cogeneration, 125 MW from biogas, 53 MW from waste-to-energy, 3 MW from solar PV plants, and 1 MW from hybrid systems.[61]

With the recently announced JNNSM described in Chapter 4, India hopes to develop more of its solar resource potential. As of June 2010, solar PV plants in India had reached a cumulative generation capacity of approximately 15.2 MW. This is approximately 0.07% of JNNSM's 2022 target of 22 GW.[62] As reported by *CSP Today*, JNNSM's goal would "make India the producer of almost three-quarters of the world's total solar energy output."[63]

By the end of the Tenth Plan (2007), India achieved a cumulative installed capacity of 10.161 GW of renewable energy (see Table 2-1). Additions totaling 15 GW are targeted during the Eleventh Plan to bring the total installed grid-connected renewable generating capacity to over 25 GW. Wind energy is expected to contribute approximately two-thirds of the added capacity in this plan period. If India is able to achieve its renewable energy goals by 2022 (by the end of the Thirteenth Plan), it will reach a total of 74 GW of installed capacity for wind, solar energy, biomass, and small hydropower, with wind and solar expected to account for more than 80% of the installed renewable power.

Although the government provides assistance for renewable energy implementation in the form of generation-based incentives (GBIs), subsidies, subsidized credits, and reduced import duties, the Indian market does not offer investors a framework that is as investor-friendly as in some developed countries. The main reason is that renewable energy sources are not systematically prioritized over non-renewable sources at a given national budget and a given power demand scenario. While the market certainly offers great opportunities for investors, it also requires adaptation and entrepreneurship to develop solutions that specifically fit the Indian scenario.

Off-grid applications for rural electrification and captive power for industries offer a promising opportunity for renewable energy technologies in India. Both of these applications can benefit from renewable energy's advantages over conventional energy sources: local control of the energy resource and power system and suitability to smaller-scale applications. Renewable energy's competition is typically either a costly connection to the national grid or diesel generator-based power with its high maintenance and fuel costs. On average, the cost of producing power for a coal plant is about INR 2 (USD 0.03) per kWh, while electricity from a diesel generator plant is approximately INR 10 (USD 0.20) per kWh. To compete effectively with these established technologies, renewable energy technologies require business models adapted to the characteristics of renewable power plants that include plans for efficient marketing, distribution, operation and maintenance, and access to financing.

For on-grid application of renewable energy, growth depends on grid infrastructure improvements and the continued reduction of renewable energy costs. Currently, wind, small hydro, and biomass are the most cost-competitive renewable options. Solar technologies, including concentrated solar power (CSP) and PV, are the least competitive but offer the greatest opportunity for growth because of the high potential. It therefore receives the most financial support in terms of government incentives.

2.2. Renewable Energy Application in Industrial Use and Transportation

A large percentage of renewable energy in India is covered under captive generation for industrial use. This is especially true in the wind market where 70% of electricity from wind projects is produced for direct consumption by large industrial facilities to mitigate the effect of frequent shortages of electricity from the national grid. Telecommunications companies are also looking toward renewable energy as they search for new solutions to power India's 250,000 telecom towers. Systems such as solar PV-based hybrid systems provide a less polluting alternative to diesel power, serve as a hedge against increasing diesel fuel prices, and help minimize the logistical challenges of transporting and storing diesel fuel at remote tower locations.[67]

For the last 2 years, solar cooling has been a buzzword in the industry. While its attraction in a country as sunny and hot as India is obvious, the technology is still under development and is not yet economically viable. There are, however, some demonstration sites such as the Muni Seva Ashram in Gujarat, which uses parabolic Scheffler-type dishes to supply a 100- ton air-conditioning system.[68]

On the transportation front, there have been initiatives to switch to alternative transportation fuels such as compressed natural gas and electricity. The Reva, developed by the Maini Group, is India's—and one of the world's—first commercially available electric car. TATA and General Electric are also in the process of developing electric vehicles.

Table 2-1. Development of Grid-connected Renewable Power in India (in MW)[64]

Five-year Plan	Achieved		In Process	Anticipated	Targets
	By the End of the 9th Plan (cumulative installed capacity)	10th Plan (additions during plan period)	Anticipated in the 11th Plan (additions during plan period)	By the End of the 11th Plan (cumulative installed capacity)	By the End of the 13th Plan (cumulative installed capacity)
Years	Through 2002	2002–2007	2007–2012	Through 2012	Through 2022
Wind	1,667	5,415	10,500	17,582	40,000
Small Hydro	1,438	520	1,400	3,358	6,500
Biomass	368	750	2,100	3,218	7,500
Solar	2	1	1,000	1,003	20,000[65]
Total	**3,475**	**6,686**	**15,000**	**25,161**	**74,000**

Table 2-2. Power Generation Costs in India by Energy Source (2008)

Energy Type	Electricity Generation Costs in INR/kWh (USD/kWh)[66]	Source
Coal	1–2 (0.02–0.04)	McKinsey - Powering India
Nuclear	2–3 (0.04–0.06)	McKinsey - Powering India
Large Hydro	3–4 (0.06–0.08)	McKinsey - Powering India
Gas	4–6 (0.08–0.12)	McKinsey - Powering India
Diesel	10+ (0.20+)	McKinsey - Powering India
Wind (on-shore)	3–4.5 (0.06–0.09)	Industry experts
Small Hydro	3–4 (0.06–0.08)	Industry experts
Biomass	4–5 (0.06–0.10)	Industry experts
Solar (CSP)	10–15 (0.20–0.30)	Industry experts
Solar (PV)	12–20 (0.24–0.40)	Industry experts

In addition, highly visible pilot projects are deployed to increase public interest in renewable energy technologies. The October 2010 Commonwealth Games in New Delhi are showcasing renewable energy for transportation and other uses including the utilization of at least 1,000 solar rickshaws, which use PV-powered motors for transporting athletes at the games.[69] Also, a 1 MW PV plant will provide electricity for one of the stadiums at the games.[70]

Liquid biofuels, namely ethanol and biodiesel, are considered substitutes for petroleum-derived transportation fuels. In India, ethanol is produced by the fermentation of molasses, a by-product of the sugar industry, but more advanced conversion technologies are under development, which will allow it to be made from more abundant lignocellulosic biomass resources such as forest and agricultural residues. Biodiesel production is currently very small, using non-edible oilseeds, waste oil, animal fat, and used cooking oil as feedstock. However, given the fact that India consumes more diesel than gasoline in the transportation sector, it is expected that the production of biodiesel and other biomass-derived diesel substitutes will grow over the next decade.

2.3. National Renewable Energy Institutions and Policies

In 1992, the Indian government established the world's first ministry committed solely to the development of renewable energy sources, The Ministry of Non-Conventional Energy Sources, which has since been renamed the Ministry of New and Renewable Energy. MNRE's role is to facilitate research, design, and development of new and renewable energy that can be deployed in the rural, urban, industrial, and commercial sectors.[71] MNRE undertakes policymaking, planning, and promotion of renewable energy including financial incentives, creation of industrial capacity, technology research and development, intellectual property rights, human resource development, and international relations.[72]

MNRE's mission is to reduce India's dependence on imported oil, thereby improving the country's energy security supply; to increase clean power's share of the national energy mix; to increase the existing energy supply with a focus on improving access to clean energy; and to help new and renewable energy technologies to be cost-competitive. The ministry supports both on- and off-grid power generation from renewable sources including small hydro, wind, solar, biomass, and industrial/urban wastes. MNRE also has programs focused on rural areas, some of which supply electricity to remote villages and some promote and expand solar energy applications. MNRE has established research, design, and demonstration projects in new areas such as geothermal, hydrogen energy, and fuel cells.[73] Moreover, MNRE supervises national institutions such as the Solar Energy Centre (SEC), the Centre for Wind Energy Technology (C-WET), and IREDA.

IREDA, established in 1987 as a Public Limited Government Company, extends financial assistance for renewable energy and energy efficiency projects.[74] IREDA is registered as a non-banking financial company and arranges its resources through market borrowing and lines of credit from bilateral and multilateral lending agencies such as the Asian Development Bank (ADB), World Bank, the Nordic Investment Bank, Japan International Cooperation Agency, the French development agency AFD, and the German development bank KfW. IREDA provides term financing for renewable energy and energy efficiency projects including hydro energy, wind energy, bioenergy (biomass, biofuels, and waste-to-energy) and solar energy. A number of incentive schemes of MNRE are administered through IREDA.

There are a number of government institutions whose competence extends into the renewable energy sector. The Ministry of Power (MoP) deals with the planning of power supply, provision of political guidelines, investment decisions for government projects, training of experts, administration of laws for power generation from conventional sources, and power transmission and guidelines. MoP is responsible for the implementation of the Electricity Act 2003, the Energy Conservation Act of 2001, and JNNSM.[75]

Important government units like the CEA, the Central Electricity Regulatory Commission (CERC), the Power Finance Corporation (PFC), the PTC India Ltd., and the Rural Electrification Corporation fall under the purview of MoP.

- CEA's task is to develop a suitable energy policy for India and conduct planning and coordination tasks. It also completes the preliminary analysis for MoP on technical and economic issues.[76]
- CERC was established in 1998 under the Electricity Regulation Act as an independent, central regulation authority. It defines the tariffs for the public sector power producers and advises the government in matters of tariff and competition policy. The State Electricity Regulatory Commissions (SERCs), established at the level of the federal states, control the generation and distribution markets in their respective states. They monitor the quality of the services, tariffs, and fees. [77]
- PFC is responsible for tapping new sources of finance for investments in power projects in the public and private sectors.[78]
- PTC India Ltd., formerly called Power Trading Corporation of India Limited, was established in 1999 with a mandate to optimally utilize the existing resources to develop a full-fledged, efficient, and competitive power market to attract private investment in the Indian power sector and to encourage the trade of power with neighboring countries.[79]
- The Rural Electrification Corporation Ltd, established in 1969, is responsible for the financial support of all rural electrification programs including the large-scale *Rajiv*

Gandhi Grameen Vidyutikaran Yojana (RGGVY), which aims to extend electricity to all rural households and households below the poverty line.[80]

The central government's renewable energy goals are included in several planning and policy documents. For example, the Planning Commission, established in 1950, is responsible for formulating five-year plans that prioritize the development and use of national resources. The Prime Minister's Council on Climate Change oversees India's NAPCC and periodically reviews key policy decisions affecting climate change. These plans are relevant for the renewable energy sector at the long-term, national policy level.

In addition to the cross-technology plans discussed in Sections 2.3.1–2.3.5, the Government of India has enacted technology-specific policies to support renewable energy deployment. These technology-specific policies are described in Sections 3–7.

2.3.1. Eleventh Plan

India's five-year plans are developed under the leadership of the Planning Commission and describe a strategic plan of activities to be undertaken by the Government of India in support of priority objectives within the resources available. The Tenth Plan period ended in 2007, and the currently active Eleventh Plan ends in 2012. The Planning Commission hired a new and renewable energy working group to develop plans for renewable energy during the Eleventh Plan and prospective plans for the Twelfth and Thirteenth Plans. The group also looked into energy access programs using renewable technologies and identified necessary fiscal policies to support these programs and plans. As a result of the working group's proposals, the Eleventh Plan established the goal that 10% of power generation capacity shall be from renewable sources by 2012 and represent a 4%–5% share in the electricity mix.[81] To achieve this, subsidies that promote investment without ensuring optimal performance (e.g., depreciation benefits) are to be phased out and replaced with incentives that emphasize performance (e.g., feed-in tariffs).

2.3.2. Electricity Act 2003

The Electricity Act 2003 has been a major step towards liberalizing the power market in India along the value chain, encouraging competition and attracting investment. It also mandates that the SERC establish requirements for the minimum purchase of electricity from renewable sources and allows penalties for non-compliance; SERCs are expected to use a RPO. Under the Act's Part VII Section 61(h), the promotion of cogeneration and electricity generation from renewable sources is identified as a consideration in the establishment of tariff regulations, allowing for the CERC to establish a preferential tariff for renewable energy.[82] Further, the "open access" provision allows licensed renewable energy power generators access to transmission lines and distribution systems and only requires that the generators pay a wheeling fee for use of the transmission lines and a fee to the load dispatch center. [83]

2.3.3. National Electricity Policy 2005

NEP stipulates the need for increasing the share of electricity from non-conventional sources and allows for the SERCs to establish a preferential tariff for electricity generated from renewable sources to enable them to be cost-competitive.[84] Section 5.12.3 of NEP encourages the development of cogeneration facilities and allows for SERCs to promote arrangements between cogenerators and distribution companies interested in purchasing excess electricity through a competitive bidding process.

2.3.4. National Tariff Policy 2006

The National Tariff Policy, announced in January 2006, mandates each SERC to specify a RPO with distribution companies in a time-bound manner. Again, these purchases are to be made through a competitive bidding process. The objective of this policy is to enable renewable energy technologies to compete with conventional sources. Section 6.4 of the National Tariff Policy calls for the relevant commission to establish preferential tariffs with distribution companies for the purchase of electricity from non-conventional technologies.[85]

2.3.5. Rural Electrification Programs

RGGVY was launched by the central government in 2005 with the goal of extending electricity to all rural households and to households below the poverty line.[86] Under this program, 90% of the capital costs are subsidized by the central government. As of 2009, only 44% of Indian households had been electrified.[87] To electrify some of these villages and hamlets, the government has used distributed renewable as well as non-renewable energy sources. During the Tenth Plan, renewable energy sources were tapped to electrify approximately 5,000 villages and hamlets with the majority of communities being served by solar energy. For a more detailed description, see Chapter 7.

Table 2-3. Actors in the Indian Power Market Along the Value-creation Chain[88]

	Central Government		State	Private
Policy	MoP, MNRE		State governments	-
Planning	CEA		SEB	-
Regulation	CERC		SERCs	-
Generation	National generation utilities		State generation utilities	Independent power producers
Transmission	-	Power Grid Corporation of India Limited	State transmission utilities	Increasing number of private service providers
Execution	Regional load dispatch centers		State load dispatch centers	
Distribution	-		State distribution utilities	Small number of private service providers
Trade	PTC India Ltd. Licensee		Licensee	Licensee
Law	**Appeal tribunal**			

2.4. Governance and Institutional Arrangement of Renewable Energy

The liberalization and reform process initiated in 1991 with the division of the energy ministry into MoP, the Ministry of Coal, and MNRE has led to large-scale changes in the governance structure of the power market. Today, along with the traditional central government and state-level actors, there is also an increasingly active private sector at most of the points along the value chain, especially in generation, transmission, and power trading (see Table 2-3). Private interest remains low in the distribution network.

As of yet there is no overarching renewable energy law governing all states. Instead, there are separate initiatives by the central and state governments. For instance, JNNSM is an initiative of the central government, whereas RPOs and RECs come under state jurisdiction. For specific technologies, central government policies and guidelines have been implemented to different degrees by individual states, which can result in inconsistencies between states.[89] For example, states have different policies regarding which entity (developer, power purchaser, or transmission and distribution company) is required to finance the extension of transmission and distribution lines when generation facilities are developed beyond the reach of the current grid. States also have different regulations regarding technical standards such as mandating the location of the meter, which affects the measurement of the amount of energy that is sold to the grid.[90]

2.5. Grid Connection and Status Overview

In March 2009, the Indian power network had a total length of 7.49 million circuit kilometers (ckm).[91] In comparison to the power generation sector, investments into the transmission and distribution networks have been lower in recent years. Nevertheless, the transmission network has improved considerably. The distribution network, however, remains in a poor state. In the ongoing Eleventh Plan, the high-voltage network is to be extended by around 95,000 ckm to a capacity of more than 178,000 mega volt amperes (MVA). In the low- voltage area, an additional 3,253,773 ckm and a capacity of 214,000 MVA are to be added. Another extremely important task is the "Power for All by 2012" mission,[92] declared by the Government of India—the ambitious goal of providing power to all Indian villages by 2012, to a large extent through grid access.

2.6. Renewable Purchase Obligations and Renewable Energy Credits

2.6.1. Renewable Purchase Obligations

Section 86(1)(e) of the Electricity Act 2003 requires that the state commissions specify RPOs for the obligated entities. It says, "The State Commission shall discharge the following functions, namely: ... promote cogeneration and generation of electricity from renewable sources of energy by providing suitable measures for connectivity with the grid and sale of electricity to any person, and also specify, for purchase of electricity from such sources, a percentage of the total consumption of electricity in the area of a distribution license."[93] In pursuance of this policy, the National Tariff Policy mandates that each SERC specify RPOs by distribution licensees in a time-bound manner. As of April 2010, 18 states have established RPOs or have draft regulations under consideration (see Table 2-4) with RPO requirements ranging from 1% to 15% of total electricity generation.

Table 2-4. State RPOs[94]

State	2009–2010 RPO	Out Year RPO (if designated)	Notes
Andhra Pradesh	5%		5% in each year from 2004–2008, no RPO requirements specified thereafter
Assam	5%	2015–2016 15%	Draft regulation currently under consideration
Chhattisgarh	Biomass 5%, Small Hydro 3%, Others 2%		Same percentages for each year from 2007–2010, no RPO requirements specified thereafter
Delhi	1%		1% RPO for four separate utilities (NDPL, BYPL, BRPL, NDMC) from 2006–2010, no RPO requirements specified thereafter
Gujarat	2%	2012–2013 7%	
Haryana	10%	2010–2011 10%	
Jharkhand			Draft regulation under consideration to establish an RPO of 5% for each year from 2010–2013
Karnataka	7%–10%	2010–2011 7%-10%	The RPO percentage depends on the distribution company with a maximum set at 20%
Kerala	Wind 2%, Small Hydro 2%, Others 1%		5% each year from 2006–2009, no RPO requirements specified thereafter
Madhya Pradesh	Wind 6%, Biomass 2%, Cogen 2%, Others 2%	2011–2012 12%	
Maharashtra	5%		RPO specified from 2005–2009, no RPO requirements specified thereafter; draft regulation currently under consideration to increase the RPO from 6% in 2010–2011 to 10% in 2015–2016 and to include a solar set aside
Orissa	5%	2015–2016 8% (0.5% increase/year)	Draft regulation currently under consideration
Punjab	3%	2010–2011 4%	
Rajasthan	Wind 6.75%, Biomass 1.75%, Others 8.5%	2010–2011 Wind 7.5%, Biomass 2%, Others 9.5%	
Tamil Nadu	13%	2010–2011 14%	
Uttarakhand	9%	2010–2011 10%	
West Bengal	7%–9%	2010–2011 10% for all utilities	RPO depends on the utility until 2010 when all are required to meet the 10% goal

2.6.2. Tradable Renewable Energy Credits

Naturally, the availability of renewable energy sources differs across India. In some states, such as Delhi, the potential for harnessing renewable energy compared to the demand for energy is very small. In other states, such as Tamil Nadu for wind, Rajasthan for solar, or Himachal Pradesh for hydro, it is very high. This offers opportunities for inter-state trading in the form of RECs.

Such trade allows for more economically efficient development of renewable energy throughout the country as distribution licensees in states with limited resources can purchase RECs associated with renewable generation in other states where it is less expensive to develop renewable energy projects. In this way, each state's RPO can be met in the most economically efficient manner. In January 2010, CERC announced the terms and conditions for a tradable REC program as follows:

- "There will be a central agency, to be designated by CERC, for registering RE generators participating in the scheme.
- The renewable energy generators will have two options [Shown in Figure 2-1]: either sell the renewable energy at a preferential tariff fixed by the concerned Electricity Regulatory Commission, or sell the electricity generation and environmental attributes associated with RE generation separately.
- On choosing the second option, the environmental attributes can be exchanged in the form of REC. Price of the electricity component would be equivalent to the weighted average power purchase cost to the distribution company, including short-term power purchase but excluding renewable power purchase cost.
- The central agency will issue the REC to renewable energy generators.
- The value of one REC will be equivalent to 1 MWh of electricity delivered to the grid from renewable energy sources.
- The REC will be exchanged only in the power exchanges approved by CERC within the band of a floor price and a forbearance (ceiling) price to be determined by CERC from time to time."[95]

CERC issued an amendment to the terms in September 2010 clarifying participation of captive generation plants and restricting participation of any generator terminating an existing PPA to sell power under the REC scheme.[96] The two paths under which renewable power will be sold under the REC program are illustrated in Figure 2-1.

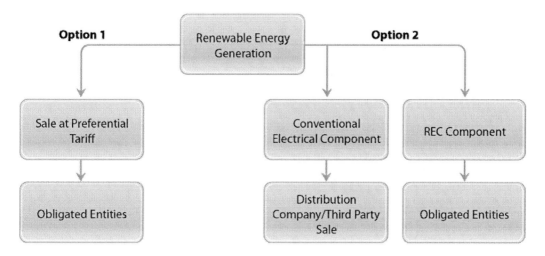

Figure 2-1. Route for sale of renewable energy generation.

3. WIND POWER

Section Overview

India has been a pioneer in the commercial use of wind energy in Asia since the 1990s. In 2009, India had the fifth largest installed wind capacity globally, only behind the United States, China, Germany, and Spain. During that year, India added 1,338 MW[97] of wind capacity for a total installed capacity of 10,925 MW. This represented a 14% annual growth rate and contributed 3.5% to the global wind market.[98] The most recent data available at the time of this writing show that India's wind capacity totalled 12,009 MW at the end of June 2010, which represented 70% of India's total renewable energy capacity.[99] India's robust domestic market has transformed the Indian wind industry into a significant global player.

The success of the Indian wind market can be attributed to the quality of the wind resource and to government incentives, which became available early on as the global wind industry began to grow. Indian company Suzlon is the market-leader in wind power in Asia and the third largest manufacturer of wind turbines in the world. In combination with its German subsidiary REpower, Suzlon has a world market share of 12.3% in installed new capacity.[100]

3.1. Wind Power Potential

The C-WET resource assessment program has estimated the potential for wind installation for nine states, as shown in Table 3-1. The estimates are based on the assumption that 1% of each state's land area is available for development and that each megawatt of wind capacity requires 12 ha of land. The assessment shows that India's total wind potential is 48,561 MW, with Karnataka, Gujarat, and Andhra Pradesh as the leading states.[101]

Table 3-1. Wind Potential for Nine Indian States (C-WET)[102]

State	Wind Power Potential (MW)
Andhra Pradesh	8,968
Gujarat	10,645
Karnataka	11,531
Kerala	1,171
Madhya Pradesh	1,019
Maharashtra	4,584
Orissa	255
Rajasthan	4,858
Tamil Nadu	5,530
Total	**48,561**

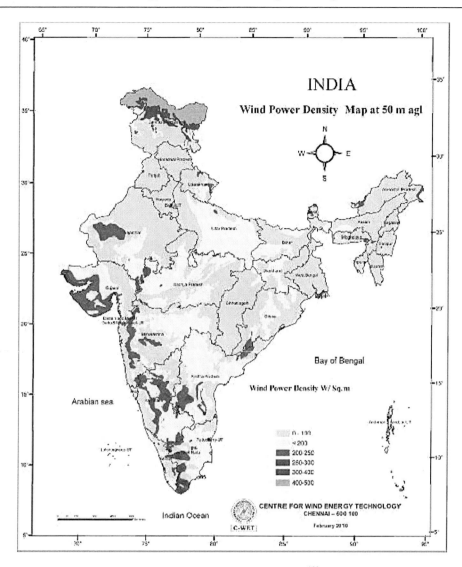

Figure 3-1. Wind power density at 50 m above the ground for India[103].

C-WET has also published a map of wind resource distribution across India (see Figure 3-1). This map shows large areas suitable for wind power development with estimated wind power density greater than 200 W/m^2 at a height of 50 m above the ground. Further assessment in the north, which the map shows as having exceptionally good resource, could substantially increase the total estimated wind potential for India, although projects in these parts could be constrained by additional development costs due to the remoteness of these areas. There has been no systematic assessment of offshore wind power potential to date.

The capacity potential of 48,561 MW estimated by C-WET and officially accepted by MNRE is lower than those of industry associations. The Indian Wind Turbine Manufacturers Association argues that by looking at higher heights above the ground to match 55–75 m turbine hub heights of the latest turbines and taking into account higher conversion efficiencies due to recent improvements in wind technology, the wind potential in India is closer to between 65,000 and 70,000 MW.[104] The Global Wind Energy Council (GWEC) outlines three different wind growth scenarios and their associated installed capacity estimates for 2030: reference, moderate, and advanced. The reference scenario follows the International Energy Agency (IEA) World Energy Outlook 2007, the moderate scenario takes into account all policy actions in planning stages or beyond, and the advanced scenario assumes that all industry-supported policy actions are implemented. All three GWEC scenarios assume smaller wind turbines will be replaced with modern models, average turbine capacities will increase, capacity factors will increase from improved technologies and sitings, and capital costs will decrease. The advanced scenario estimate is much higher than any of the other estimates at 241,349 MW installed by 2030, the moderate scenario is 142,219 MW, and the reference scenario estimate is 27,325 MW.[105]

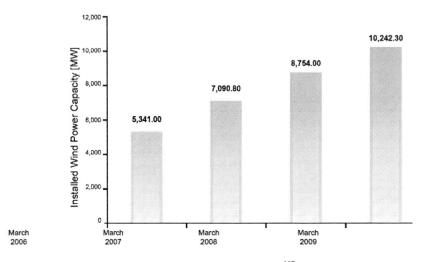

Figure 3-2. Growth of installed wind power capacity in India[107].

Table 3-2. Wind Capacity Additions by State

State	Total Installed Wind Capacity 2007 (MW)[110]	Total Installed Wind Capacity 2009 (MW)[111]	Wind Capacity Additions 2008–2009 (MW)
Andhra Pradesh	123	123	0
Gujarat	875	1,712	837
Karnataka	917	1,391	474
Kerala	2	27	25
Madhya Pradesh	70	213	143
Maharashtra	1,646	2,004	358
Rajasthan	496	855	359
Tamil Nadu	3,712	4,596	884
Others	4	4	0
Total	**7,845**	**10,925**	**3,080**

3.2. Installed Capacity and Power Generation

In March 2006, India had approximately 5,300 MW of installed wind capacity. Between 2006 and 2009, the annual rate of wind capacity additions decreased, as shown in Figure 3-

2.[106] In certain states, however, as new markets opened up, growth rates were above average. For example, between 2008 and 2009, annual growth rates of installations were 37% in Rajasthan, 33% in Karnataka, and 25% in Gujarat, while the average across India was 17%.

India's capacity additions of more than 3,000 MW between 2007 and 2009 are shown by state in Table 3-2. The largest gains were seen in Tamil Nadu (884 MW) and Gujarat (837 MW). Tamil Nadu in south India has set itself apart from the other states with a total installed capacity of 4,596 MW, or 42% of India's total installed capacity, as of the end of 2009.[108] At the end of June 2010, India's installed wind capacity totalled 12,009 MW, representing 70% of India's total renewable energy capacity.[109]

In terms of electricity supply, in 2007, wind projects supplied 11,653 GWh to Indian consumers, representing a 1.45% share of the total electricity produced in India that year.[112] Based on the 2007 installed capacity of 7,845 MW shown in Table 3-2, this equates to a capacity factor of approximately 17.0% (11,653 GWh/yr ÷ 7.845 GW ÷ 8760 h/yr × 100%). This capacity factor is a rough indication of the average conversion efficiency of wind power plants: higher capacity factors represent more efficient plants. For comparison, India's average capacity factor in 2007 was slightly higher than China's in 2007 at 16.6% but significantly lower than Germany (20.4%), Spain (20.7%), and the United States (23.5%).[113]

3.3. Existing Policies, Measures, and Local Regulations

One provision of the Electricity Act of 2003 allowed turbine manufacturers to provide complete build-operate-manage wind solutions. As manufacturers, their primary interest was in installing as many new wind projects as possible to maximize turbine sales rather than in maximizing electricity generation of individual wind plants. Many other investors made similar decisions, choosing to install as much capacity as possible to profit from tax-depreciation benefits with little regard to how well their wind plants would perform over the long term.

Newer policies, such as GBIs and RPOs, encourage independent power producers and private investors to establish large-scale, commercial wind plants that enable wind to be a more significant part of the power mix. Based on experiences in other countries, both GBIs and RPOs are generally considered to be positive steps towards encouraging the development of wind power. These can lead to developer investment in more comprehensive resource assessments at the project site and a more optimized plant design. More independent power producers may enter India's wind market, turbine sizes and tower heights are increasing, and turbine technology is continuing to improve, all of which may contribute to higher capacity factors for future wind plants in India. However, until there is a long-term guarantee for GBIs beyond 2012 and an enforcement of the RPO targets, there is unlikely to be real movement in the market.[114]

A project developer can currently choose between the central government's accelerated depreciation or GBIs; the preference cannot be changed once the choice has been made.[115]

3.3.1. State-level Renewable Energy Preferential Tariffs

In 2003, SERCs were given the right to set preferential renewable energy tariffs based on guidelines stipulated by CERC. Although several states have followed through, these tariffs are insufficient to drive further development in wind. However, the state-level GBIs can be used in conjunction with the central-level incentives (either accelerated depreciation or GBIs) to make projects more financially attractive to developers.[116]

For 2009–20 10, CERC established a preferential tariff band for wind energy projects ranging from INR 3.75/kWh (USD 0.68/kWh) to INR 5.63/kWh (USD 1.01/kWh), depending on the project size.[117] State wind energy preferential tariffs range from INR 3.14/kWh (USD 0.57/kWh) in Kerala to INR 4.08/kWh (USD 0.73/kWh) in Haryana.[118] Other states that have established feed-in tariffs include Maharashtra, Andhra Pradesh, Madhya Pradesh, Karnataka, Tamil Nadu, and West Bengal; all states have caps ranging from 50 MW to 500 MW.[119] The tariff conditions vary by state as well; some are fixed for a set number of years while the tariffs in other states are reduced or escalated during those years.

3.3.2. Accelerated Depreciation

In the past, the main incentive for developing wind power projects was accelerated depreciation. This tax benefit allows projects to deduct up to 80% of the value of wind power equipment during the first year of the project's operation. Investors are then given tax exemption status for up to 10 years.[120] Wind power producers receiving accelerated depreciation benefits must register with and provide generation data to IREDA and are not eligible to receive the more recent GBIs.[121]

3.3.3. Indirect Tax Benefits

Indirect tax benefits included exemptions or concessions on the excise duty and a reduction in customs duty for certain wind power equipment. [122] Wind-powered electricity generators and components, as well as water-pumping wind mills, wind aero-generators, and battery chargers, are exempt from excise duties. Indirect tax benefits for manufacturers of specific wind energy parts vary from 5%–25% depending on the component.[123]

3.3.4. Central-level Generation-based Incentives

Offered by the central government since June 2008 and administered by IREDA, the GBI for wind is available for independent power producers with a minimum installed capacity of 5 MW. As of December 2009, the GBI is set at INR 0.50/kWh (USD 0.01/kWh) of grid-connected electricity for a minimum of 4 years and a maximum of 10 years, up to a maximum of INR 6.2 million (USD 140,000) per MW. The scheme will deploy a total of INR 3.8 billion (USD 81 million) until 2012 and aims to incentivize capacity additions of 4,000 MW. Wind power producers receiving a GBI must register with and provide generation data to IREDA.[124] The GBI is offered in addition to SERC's state preferential renewable energy tariffs. However, IPPs using GBIs cannot also take advantage of accelerated depreciation benefits.[125] The GBI program will be reviewed at the end of the Eleventh Plan and revised as deemed appropriate. As of October 2010, 30 projects had been registered under this scheme with over 200 MW commissioned.[126]

3.3.5. Renewable Purchase Obligations

As discussed in Chapter 2, several states have implemented RPOs with a requirement that renewable energy supplies between 1% and 15% of total electricity. The impact of the RPOs on wind development may depend on the penalties and enforcement of the targets as well as an effective REC market to promote development of areas of the country with the most abundant wind resources.

3.4. Investment Flows and Industrial Trends

The cost of producing wind turbines is lower in India than in other countries as a result of lower labor and production costs relative to other countries producing wind power generation equipment. In 2009, the average installed cost for a wind power project in India was INR 53.3 million (USD 1.1 million) per MW, and the installation costs per megawatt will likely continue to decrease. By comparison, the average installation cost in the United States in 2009 was approximately INR 105.0 million (USD 2.1 million) per MW.[127]

India's wind turbine manufacturers currently produce 3,000–3,500 MW of new turbines a year. Five key players (Suzlon, Enercon, RBB Energy, NEPC, and Vestas) cover 80% of this, and they are also responsible for 90% of India's new capacity installations in 2009–2010.[128] By 2015, India's total wind turbine production is expected to rise to 5,000 MW per annum. Several new manufacturers, many of which are local, are now entering the market, and prices for turbines are expected to come down due to increasing competition. According to MNRE, India's export of wind turbines and blades to the United States, Europe, South America, and Asia is valued at INR 45 billion (USD 900 million) and imports reached INR 22 billion (USD 440 million).[129]

3.5. Technology Developments

The current wind technology developed in India has lagged behind technologies produced around the globe. According to MNRE, turbines in India range in capacity from 250 kW to 2,100 kW compared to a global maximum of 5,000 kW; hub heights range from 41 m to 88 m compared to a global maximum of 117 m; and rotor diameters range from 28 m to 80 m, compared to a global maximum of 126 m.[130]

The average capacity of installed wind turbines in India was only 400 kW in 2000; by 2008 it had reached 1 MW. As plant sizes increase, installation costs decrease. GWEC expects that India's average turbine size will reach 1.5 MW by 2013.[131]

Larger turbines require less land for an equal installed capacity, reduced operation and maintenance, and faster commissioning, and very large turbines are starting to enter the market. For example, Suzlon is now assembling turbines of 5 MW and 6 MW through its German subsidiary, REpower.[132]

Upcoming technological developments include wind forecasting to enable integrated grid management and more efficient generation. [133] There is also a rising interest in offshore wind developments for India, although there has not yet been any significant progress.[134]

3.6. Local Case Studies

A major success story in the Indian renewable energy sector has been Suzlon, a company set up in 1995 by Tulsi Tanti. By 2009, Suzlon was the third largest wind turbine manufacturer in the world with a market share of 9.8%.[135] Suzlon is the leading wind power company in India with a domestic market share of more than 44% of installed capacity and 26% of wind turbines sold.[136] Its success has been attributed to fast delivery and turnaround times that undercut the competition.[137] In 2006, Suzlon bought the Belgian gearbox maker Hansen, and in 2007, it bought REpower, a German company with offshore wind experience. The company has suffered some recent setbacks, including damage to some of its blades under different environmental conditions. Aggressive leverage for investments, costly blade repair programs, and a global recession have caused the company some recent financial trouble.[138]

In June 2010, Suzlon reported a loss of INR 1.8 billion (USD 36 million), citing an order backlog.[139]

Bharat Heavy Electricals Ltd (BHEL), one of India's largest power equipment manufacturers, is entering the wind turbine market in 2010 to help meet the demand for larger turbines in India. The company plans to develop their own blade manufacturing facilities in India, but they plan to have towers manufactured in India and blades sourced from foreign manufacturers in the near term as they look to support deployment of 1.5–2.0 MW turbines. BHEL was an early manufacturer of 250 kW wind turbines but ceased production in response to a drop in demand.[140]

On the project development side, TATA Power is planning to expand its wind capacity from 200 MW to 500 MW by 2013 and to 2,000 MW by 2017.[141] Airvoice has announced plans to set up 3,000 MW of wind in Karnataka with 200 MW to be installed in the first phase.[142] NHPC Limited, a government enterprise and the largest hydropower operator of the country, is currently building a 100 MW wind plant in Madhya Pradesh.[143] Another government enterprise, Neyveli Lignite Corporation Ltd, announced a 50 MW project in the south-Indian harbor city of Tuticorin[144] to supply power to a lignite (brown coal) open-pit mining network currently under construction. Oil and Natural Gas Corporation Ltd also has plans to invest INR 65 million (USD 123 million) in a 50 MW project in Gujarat as well as in smaller wind power plants in Karnataka.[145] The company has plans to install up to 2,000 MW of wind capacity, mainly in the state of Gujarat.[146]

3.7. Success Stories from the International Community and Potential Opportunities for India

India's wind industry continues to offer attractive opportunities for investors. In addition to capacity growth through new project development, many outdated and smaller plants will be re-powered, where outdated lower capacity turbines are replaced by modern and larger equipment.[147]

Foreign companies are also becoming involved in the manufacturing sector. For example, Siemens has announced plans to invest INR 5 billion (USD 100 million) to open a 200 MW wind turbine production plant in India with plans to scale up to 500 MW over 3 years. The

Spanish manufacturer Gamesa has already commenced operations at its 200 MW manufacturing facility for 850 kW turbines in Tamil Nadu.[148]

With 7,600 km of coast, the installation of offshore wind plants can play an important role in the growth of India's wind market. Although there is currently little information available regarding the potential for offshore wind in India, C-WET released data in 2008 showing modest prospects along the western coastline with several sites prone to cyclonic conditions. Data collected at the southern coastline, at a monitoring site in Rameshwaram, indicates some potential. However, much of the data has been extrapolated and further in-depth assessment is required.[149] So far, there has also been little progress by the private sector into exploring this potential.[150] Companies such as Areva, Siemens, General Electric, and Suzlon are now exploring opportunities in offshore wind in India as the conditions for low-cost generation are improving.[151]

4. SOLAR POWER

Section Overview[152]

Solar is an important, although currently underutilized, energy resource in India with the potential to offer an improved power supply (especially in remote areas) and increase the security of India's energy supply. On average, the country has 300 sunny days per year and receives an average hourly radiation of 200 MW/km^2. The India Energy Portal estimates that around 12.5% of India's land mass, or 413,000 km^2, could be used for harnessing solar energy.[153] This area could be further increased by the use of building-integrated PV. Though large-scale CSP has not yet been deployed in India, one study has estimated that this technology alone could generate 11,000 TWh per year for India. In addition to India's potential for widespread deployment of solar technologies to supply electricity across the country, India also has the potential to significantly reduce electricity demand through increased deployment of solar water heaters (SWH), which can be deployed on rooftops in the built environment.

Although India already has a strong solar cell production industry, until now, there has not been a high demand for them in the domestic market. In response to the announcement of JNNSM in November 2009, substantial expansion in the domestic solar market is anticipated. JNNSM's target of achieving at least 20,000 MW[154] of grid-connected solar power by 2022 could make India one of the leading solar countries in the world, not only in total installed solar capacity but also in manufacturing components and technology research and development. The target encourages development of both PV and CSP technologies by allocating 10,000 MW of the goal to each technology.

According to JNNSM, a further 2,000 MW of off-grid solar power capacity are to be installed by 2022. In order to achieve the goals, MNRE seeks to create an attractive environment for investors, including incentives such as feed-in tariffs.[155,156] Policies enacted for the first phase will be reviewed in the summer of 2011 to incorporate first lessons learned.

4.1. Solar Photovoltaics

India's installed solar power capacity of 15.2 MW at the end of June 2010 was based entirely on PV technology with approximately 20% of the capacity being used for off-grid applications.[157] Currently, more attention is being paid to large-scale solar PV projects. In Phase 1 of JNNSM, which ends in 2013, India aims to install 500 MW of grid-connected solar PV power.[158] New PV projects are also being registered under state programs such as in Punjab, Gujarat, West Bengal, Rajasthan, and Karnataka, though many of these are being migrated to JNNSM. The creation of special economic zones that provide land, water, and power as well as financial incentives has spurred growth in the domestic manufacturing sector. International companies from all over the world are now lining up to get a share of India's solar market, which is valued at INR 3,500 billion (USD 70 billion) through the end of JNNSM in 2022.

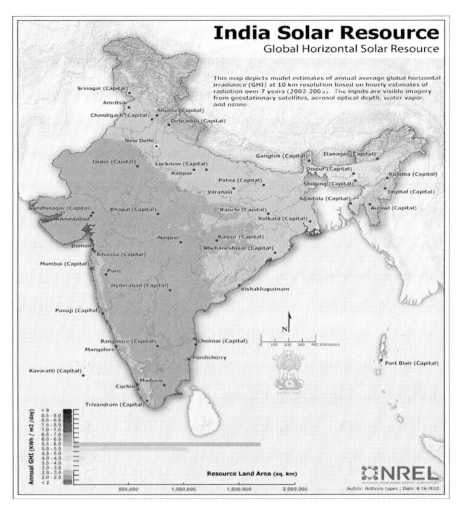

Figure 4-1. India's GHI resource at 10 km resolution.

4.1.1. Resource/Technological Potential

Estimates of global horizontal irradiance (GHI) for India are available from several different organizations. The U.S. National Aeronautics and Space Administration (NASA) Surface Meteorology and Solar Energy (S SE) dataset provides estimates of GHI at a 1° spatial resolution, which is approximately 100 km at 26° N latitude and can be visualized and downloaded from the Solar and Wind Energy Resource Assessment (SWERA) Web site.[159] Solar resource maps at 40-km spatial resolution developed using the National Renewable Energy Laboratory (NREL) Climatological Solar Radiation model are also available from SWERA.

NREL has recently released 10-km resolution solar resource maps for India based on the SUNY satellite to irradiance model.[160] Figure 4-1 shows the annual average GHI across India, which illustrates that most areas of the country have greater than 5.5 kWh/m^2/day of GHI, and the amount reaching a collector could be augmented by optimizing its orientation.

4.1.2. Existing Grid-connected and Off-grid Installed Solar Photovoltaics Capacity

As of June 2010, India's grid-tied PV solar power was only 12.28 MW, or 0.07%, of the total 17,174 MW of grid-connected renewable power capacity.[161] During 2009–2010, Solar PV projects accounted for 0.5% of grid-connected renewable energy additions.[162]

Off-grid and decentralized solar applications have been relatively more successful due to direct subsidies and government-financed pilot projects. The total decentralized installed solar capacity as of June 2010 was 2.92 MW, or 0.7%, of the total 420 MW of off-grid renewable power. [163] These decentralized applications include solar street and home lighting systems, solar lanterns, solar cookers, and water pumps.

4.1.3. Policies, Measures, and Local Regulations

Over the last decade, a number of programs administered by the central government and some state governments have tried to initiate solar power projects, albeit without much success.

The Tenth Plan (2002–2007) targeted the installation of 5 MW of solar PV; however, only 1 MW was installed during that period.[164] The Eleventh Plan (2008–2012) originally established a combined target of 50 MW[165] for both solar PV and CSP, and, in spite of an INR 15 (USD 0.30) feed-in tariff over 12 years offered by MNRE, no new projects were built due to the relatively high cost. India's current federal and state policies designed to encourage deployment of grid-tied solar PV are discussed in Sections 4.1.3.1 and 4.1.3.2, some of which also promote off-grid PV. Policies focused exclusively on off-grid renewable systems are covered in the decentralized energy chapter (Chapter 7).

4.1.3.1 JNNSM

Officially launched in November 2009, JNNSM is one of the eight National Missions laid out in India's NAPCC. It aims to incentivize the installation of 22,000 MW of on- and off-grid solar power using both PV and CSP technologies by 2022 as well as a large number of other solar applications such as solar lighting, heating, and water pumps. As the power trading arm of the National Thermal Power Corporation (NTPC), NTPC Vidyut Vyapar Nigam Ltd (NVVN) has been designated as the nodal agency to ensure the execution of Phase 1 of the mission.[166]

JNNSM aims to address the shortcomings of prior schemes through revised and more attractive feed-in tariffs, a single-window application process, and RPOs that include a solar purchase obligation.[167] In February 2010, CERC announced a feed-in tariff for financial year 2010–2011 of INR 17.9 (USD 0.36) per kWh for PV and INR 15.3 (USD 0.31) per kWh for CSP and declared that Power Purchase Agreements (PPAs) would have a validity of 25 years. It is assumed that at current cost levels, the tariff will allow investors to achieve an internal rate of return of about 16%–17% after taxes. CERC will revise the tariff every year. Ideally, by 2022, installation costs will come down significantly to enable solar power to achieve grid parity so that it becomes a viable source for India's energy needs in the absence of government incentives. Assuming a continuing decrease in PV costs over the span of JNNSM, even as the preferential feed-in tariff is reduced in subsequent years, the share of solar power in the energy mix should continue to increase.[168] Under JNNSM, the NVVN is required to purchase the expensive solar power from developers and bundle it with an equivalent amount of its much cheaper coal-based power before selling the mixed power to the various utilities at a marketable price.[169]

Three phases are identified under JNNSM. Grid-connected solar projects that signed PPAs prior to November 19, 2009, were eligible to migrate to JNNSM under certain conditions until February 29, 2010.[170] At the end of each phase there will be a thorough reevaluation of the process.[171]

Phase 1: 2010–2013

The focus during Phase 1 is to experiment with incentive structures and to create a market for solar power in India by bringing in investors, engineer-procure-construct (EPC) contractors, and equipment manufacturers. Both on- and off-grid projects will be promoted during this phase with the expectation that 500 MW of grid-connected and 200 MW of off-grid solar PV will be installed.[172]

The allocation of 500 MW of grid-connected PV projects will be decided in two batches. The first will be in financial year 2010–2011 and the second in financial year 2011–2012. The first batch will allocate 150 MW. Under migration guidelines, projects migrating from older incentive schemes to newer ones should be selected prior to new projects. If applications exceed 150 MW, projects will be chosen based on the discount offered by project developers on the CERC tariff. Under the second batch, additional projects will be selected up to 350 MW of remaining capacity. Projects will feed into the grid at 33 kV or above. Individual projects will have a capacity maximum of 5 MW (± 5%), and each company can only apply for one PV project under Phase 1 (but can also apply for one CSP project).[173]

Applying for a project under Phase 1 of JNNSM requires a company to have an audited net worth of at least INR 30 million (USD 600,000) per MW of a project's installed capacity or INR 150 million (USD 3 million) for a 5 MW project in at least one of the last four financial years. A company must provide a "Bid Bond," or third-party guarantee, per megawatt for any discount on the offered tariff. The higher the discount, the higher the amount of the bond, ranging between INR 10,000 (USD 200) and INR 50,000 (USD 1,000) per MW on a graded scale. Furthermore, the company must provide an earnest money deposit (EMD) in the form of a bank guarantee of INR 2 million (USD 40,000) per MW along with the initial request for selection and, later, a performance bank guarantee of INR 3 million (USD 60,000) per MW at the time of signing the PPA. A project shall achieve financial closure within 180 days before the signing and must be commissioned within 12 months after the signing of the PPA.[174] To ensure PV module quality, modules proposed for the project must qualify to the latest edition of the following International Electrotechnical Commission (IEC) PV module qualification test or equivalent from the Bureau of Indian Standards: for crystalline silicon solar cell modules—IEC 61215; for thin-film modules—IEC 61646; and for concentrator PV modules—IEC 62108. For the first batch of Phase 1, it will be mandatory for projects using crystalline silicon technology to use modules manufactured in India.[175]

Phase 2: 2013–2017
During Phase 2, the goal is to build on the experience of Phase 1 to facilitate a substantial increase in capacity additions, significantly bring down the cost per kilowatt-hour, and achieve additional installations of 3,000–10,000 MW of combined PV and CSP capacity. JNNSM identifies the need for international support in the form of technology transfer and financial assistance in order to meet the higher goal. The central government will work to create a favorable environment for solar manufacturing, particularly for solar thermal technology manufacturing. By 2017, the goal is for the installation of 15 million m^2 of solar thermal collector area and for off-grid solar capacity to reach 1,000 MW. For Phase 2, it will be mandatory to use cells and modules manufactured in India.[176]

Phase 3: 2017–2022
Solar power is expected to achieve grid parity by 2022 through the final goals of the JNNSM: off-grid solar capacity installations will reach 2,000 MW, on-grid capacity will reach 20,000 MW, 20 million m^2 of solar thermal collector area will be installed, and 20 million solar lighting systems will be deployed in rural households.

4.1.3.2. State Policies

Independent of these national efforts, states are promoting solar power. Gujarat, for example, is promoting the installation of 350 MW solar PV by 2011. It offers a feed-in tariff of INR 15 (USD 0.30) per kWh for the first 12 years and INR 5 (USD 0.10) per kWh for the following 13 years. In addition, there is an accelerated depreciation option of 80% in the first year of operation. The program has some uncertainties, however. First, it is unclear if the accelerated depreciation falls under jurisdiction of the state or the central government. Second, Gujarat Urja Vikas Nigam Limited, the organization responsible for the bulk purchase and sale of electricity, will buy the power from the solar projects only based on its power demand. The policy process is still evolving.[177]

Allocations for the Gujarat 350 MW PV program have been made to a number of interested promoters, but actual project execution has so far lagged behind. Narenda Modi, Chief Minister of Gujarat, is driving the policy by making use of the state's land availability and excellent irradiation and is making Gujarat the Indian hub of solar technology. In addition to the state of Gujarat, other states such as Rajasthan, Maharashtra, Andhra Pradesh, and Karnataka are developing their own solar policies.[178]

4.1.4. Investment Flows and Industrial Trends

As of May 2010, there were four major PV plants connected to the grid in India. The first was a 1 MW plant by West Bengal Renewable Energy Development Corporation. The second was a 1 MW plant by private developer Azure Power in Punjab. The most recent additions are a 3 MW plant by Karnataka Power Trading Corporation and a 1 MW plant by Maharashtra Generation Corporation.[179]

Several new PV power plants are in the pipeline, including 365 MW in Gujarat and 36 MW in Rajasthan.[180] Titan Energy, a leading Indian module manufacturer and EPC contractor, and Enfinity, a global project developer, announced plans to install 1,000 MW of solar PV on 3,000 acres in Andhra Pradesh.[181]

In the past couple of years, the number of PV module and cell suppliers in India has tripled to 30.[182] As of 2009, total module production rose from less than 60 MW to over 1,000 MW per annum. Thus far, it has catered primarily to the export market.[183] Production is projected to grow at a rate of 20%–25%, reaching 2,575 MW by 2015.[184] JNNSM targets 4,000–5,000 MW in annual domestic module manufacturing capability by 2022.[185]

PLG Ltd., one of India's largest module manufacturing companies, is in the process of upgrading its 50 MW facility to 100 MW.[186] Tata BP Solar recently signed an agreement with Calyon Bank and BNP Paribas to raise INR 3,900 million (USD 78 million) to fund further development.[187] It currently has a cell manufacturing capacity of 84 MW and a module manufacturing capacity of 125 MW with plans to expand to 300 MW.[188]

The creation of special economic zones, providing land, water, and power as well as financial incentives, has further spurred growth in the domestic manufacturing sector. A Special Incentive Package Scheme (SIPS), introduced in 2007, provides subsidies of 20%–25% for semi-conductor and eco-system (solar cells and PV) manufacturing units. These financial incentives enable vertically integrated manufacturing facilities to achieve economies of scale. As of the scheme deadline of March 31, 2010, there had been 26 applications for solar PV facilities with a cumulative investment of INR 2,290 billion (USD 45.8 billion).[189]

A number of companies that have submitted proposals for new production capacities under SIPS have been approved in principal. Approved companies include Titan Energy Systems, Reliance Energies, TATA BP Solar, PV Technologies India, KSK Surya PV Ventures, Signet Solar, Indo-Solar, Solar Semiconductors, TF Solar Power, Lanco Solar, EPV Solar, and Bhaskar Solar.[190] Companies undertaking projects without a SIPS incentive include Bharat Electronics and BHEL, who together are planning a joint 250 MW PV production facility, and ONGC, who is planning a 60 MW production facility.[191]

4.1.5. Technology Development and Transfer

While production of PV modules is on the rise in India, module manufacturers are highly dependent on imports of the silicon wafers used in module manufacturing.[192] For this reason, the government wants to promote the production of silicon material in India. The Eleventh Plan proposed funding of INR 1.2 billion (USD 24 million) from the National Energy Fund to promote the manufacturing of polysilicon.[193] MNRE and MoP plan to make it mandatory for developers to supply crystalline silicon-based modules from domestic manufacturers. However, the import of solar cells to manufacture modules is still permitted.[194]

Thin-film modules are also manufactured in India. These are less expensive than crystalline modules but also less efficient.[195] Signet Solar, a U.S. company, is planning a 300 MW production facility for thin-film modules in Tamil Nadu.[196] The Indian company TF Solar Power wants to invest INR 23.5 billion (USD 470 million) in the production of thin-film modules.[197] Poseidon Solar, an Indian-based silicon recycling company, recently announced their capability to recycle Cadmium-Telluride based thin-film modules, which will significantly reduce component costs for developers.[198]

Multi-junction cells are a promising technology on the horizon; however, there has not yet been much research and development on this in India.[199]

4.1.6. Local Case Studies

In 2009, Azure Power set up the first private grid-connected solar PV plant in Punjab with a capacity of 1 MW and servicing 20,000 households.[200] In line with its ambitious targets of establishing 100 MW solar power capacity, it is now developing a 15 MW plant in Gujarat in collaboration with U.S.-based company SunEdison. According to Azure, the cost of PV installation is already coming down; its first plant cost INR 190 million (USD 3.8 million) per MW; the new plant will cost 10% less at INR 170 million (USD 3.4 million) per MW. Future costs are expected to drop down by a further 10%–12%. Both plants are under the respective state GBI schemes. Both are 30-year contracts by which power will be sold at INR 15 (USD 0.30) for the first 10 years in Punjab and the first 12 years in Gujarat. For the remaining time periods, the tariff will reduce to INR 8.93 (USD 0.18) in Punjab and INR 5 (USD 0.10) in Gujarat.[201]

RIL Solar Group—part of Reliance Industries—will provide solar power to the Commonwealth Games in Delhi in 2010.[202] The company built the first 1 MW rooftop PV installation on the Thyagaraj stadium in New Delhi in less than 3 months. The plant is expected to generate 1.4 million kWh of electricity per year, and the surplus will be delivered to the distribution grid.

Moser Baer PV, launched in 2005, is one of India's leading cell and module manufacturers. The company is a subsidiary of Moser Baer, which was founded in 1983 and has since become the second largest producer of optical storage media.[203] Moser Baer PV established itself across the value chain, offering end-to-end solutions for on- and off-grid solar PV installations. It has a production capacity of 90 MW crystalline cells, 90 MW crystalline modules, and 50 MW of thin film. The company also has solar PV power plants of 15 MW approved in Gujarat and 5 MW in Rajasthan and successfully commissioned 50 projects in India and Germany in 2009.[204]

4.1.7. Success Stories from the International Community and Potential Opportunities for India

Germany has the largest solar PV power generation capacity, despite a significantly lower solar resource compared to India and many other countries. In 2009 alone, with 3,800 MW of added capacity, it accounted for more than half of the global market. Spain was the world leader in 2008 with 2,400 MW additional PV installations.[205] The high rate of new installations in these two countries is due to highly favorable regulatory environments. Both countries offer attractive long-term feed-in tariffs (20 years in Germany and 25 years in Spain) and have provided an overall conducive investor environment. Although the tariffs are reduced every year and are about to be cut by double digits in 2010, the deployment of new solar capacity is still high. One result of this is that German cell production companies are among the most successful in the world. The German company Q-Cells was the world's top cell producer in 2008 with 570 MW produced.[206]

International companies from all over the world are now lining up to get a share in India's solar market, which is valued at INR 175 billion (USD 3.5 billion) through 2013 and INR 3,500 billion (USD 70 billion) through the end of JNNSM in 2022.[207] Solar PV companies that have cost-competitive products, such as China-based Suntech and U.S.-based First Solar, are pegged to have an advantage over others.[208]

Gujarat has proved a popular destination for investment with 24 solar developers getting 365 MW of PV capacity approved as of June 2010;[209] the market is attracting both domestic and international investors. German companies Dreisatz GmbH and Mi GmbH, each have 25 MW of approved projects in Gujarat.[210] U.S.-based project developer Astonfield recently announced a partnership with Belectric for a 5 MW plant in Osiyan, Rajasthan, which is to be migrated to JNNSM.[211] The required land is 30 acres, but Astonfield has secured an additional 185 acres for future expansion of the plant to 20 MW.[212]

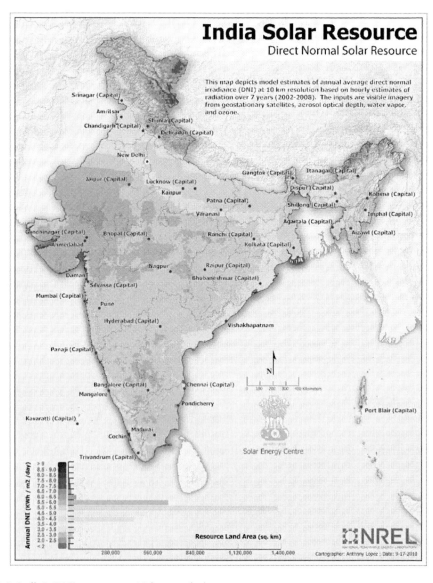

Figure 4-2. India's DNI resource at 10 km resolution.

4.2. Concentrating Solar Power

JNNSM recognizes the potential of CSP as a highly effective energy source and is expecting a significant proportion of CSP in its solar goals even in Phase 1, with 500 MW of grid-tied CSP targeted for 2013 and 10 GW targeted by 2022. CSP is currently still at the research and development (R&D) stage in India. A feed-in tariff was announced by CERC in February 2010 together with several other instruments to incentivize manufacturers for solar collector production. States such as Gujarat are introducing their own policies to stimulate CSP development, such as exempting the generated electricity from electricity duty.

4.2.1. Resource/Technological Potential

In its CSP roadmap, the IEA included northwestern India among its list of world regions showing the best solar resource for CSP projects.[213] One of the benefits that CSP offers over PV is the ability to include thermal storage in system designs. This enables a continual supply of power that does not hamper grid stability and allows CSP facilities to operate as base load plants.[214] CSP has so far been successfully implemented in only a few locations worldwide, such as the U.S. southwestern desert.[215] India's foray into CSP began in 1994 with a feasibility study for an INR 12.3 billion (USD 245 million), 35 MW CSP plant in Mathania, Rajasthan, to be co-financed by the government and KfW. However, the project was postponed indefinitely because no qualified contractors were able to submit a bid.

There are several maps and data sources available that estimate the availability of direct normal irradiance (DNI) at locations across India. The NASA SSE dataset provides estimates of DNI at a 1° spatial resolution, and NREL's Climatological Solar Radiation model provides estimates at 40-km resolution, both of which are available from SWERA.[216]

NREL has recently released 10-km resolution solar resource maps for India based on the SUNY satellite to irradiance model.[217] Figure 4-2 shows the annual average DNI across India, which illustrates that most areas of the country have greater than 5 kWh/m^2/day of DNI. This image shows the areas of highest resource occurring in the northwest desert regions and the high-elevation Himalayan region. Figure 4-2 also shows large areas with annual average DNI greater than 5.5 kWh/m^2/day in Madhya Pradesh and Maharashtra and smaller land areas with similar resources in several additional states.

A recent study by the German Aerospace Center[218] estimated the potential for CSP in several countries including India. This analysis used the NASA SSE 6.0 DNI dataset, with a spatial resolution of 100 km and remote sensing data at 1 km spatial resolution, to exclude land areas unsuitable for CSP development, including land with resource levels below 2,000 kWh/m^2/year (5.48 kWh/m^2/day), settlement areas, steeply sloping land, and agricultural areas. Table 4-1 shows estimated land area appropriate for CSP development and estimated generation potential on these lands assuming a parabolic trough system. This analysis estimates that CSP could generate almost 11,000 TWh per year in India.

4.2.2. Concentrating Solar Power Projects in Planning Stages

In the private sector, concentrating solar companies are beginning to develop a market presence in India. Areva Renewables, for example, a French company that offers CSP solutions, recently set up a partnership with the Indian companies L&T and Bharat Forge for engineering support and supply chain management.[220] Bright Source Energy, a U.S.-based solar thermal company, partnered with the French conglomerate Alstom, bringing an investment of INR 2.78 billion (USD 55 million) to the solar thermal market in India. Indian company Cargo Power and Infrastructure, a part of Cargo Motors, claimed to be the first CSP developer with a PPA in hand.[221]

Table 4-1. India: Estimated Land Area Suitable for CSP Development and Generation Potential[219]

DNI Class (kWh/m²/year)	Land Area Suitable for CSP Development (km²)	CSP Generating Potential (TWh/year)
2,000–2,099	83,522	7,893
2,100–2,199	11,510	1,140
2,200–2,299	5,310	550
2,300–2,399	7,169	774
2,400–2,499	3,783	426
2,500–2,599	107	13
2,600–2,699	976	119
2,700–2,800+	120	15
Total	**112,497**	**10,930**

Table 4-2. Approved CSP Projects in Gujarat and Rajasthan

Company	Approved CSP Project Capacity in Gujarat (MW)	Approved CSP Project Capacity in Rajasthan (MW)
Acme Telepower	46	10
Adani Power	40	
Cargo Motors	25	
Electrotherm	40	
Abengoa	40	
IDFC	10	
KG Design Services	10	
Sun Borne Energy	50	
NTPC	50	
Welspun Urja	40	
Entegra		10
Shri Rangam Brokers and Holdings		10
Total	**351**	**30**

Indian company ACME has a licensing agreement with U.S.-based eSolar, a manufacturer of modular CSP systems. Their first project, a 10 MW plant in Bikaner, Rajasthan, is in the advanced stage of installation and is expected to generate power in late 2010.[222]

NTPC, India's largest power generation company, has two CSP plants in development: a 15 MW plant in Rajasthan and a 25 MW plant in Uttar Pradesh.[223]

Approved CSP projects in Gujarat[224] and Rajasthan,[225] shown in Table 4-2, total 351 MW and 30 MW, respectively.

4.2.3. Existing Policies, Measures, and Local Regulations

In the past, there were several policy initiatives to establish CSP power generation in India, though to date there are no operational CSP plants. The failure of these initiatives is attributed to insufficient incentives and the historical absence of developers and EPC providers in the global CSP market. During the Tenth Plan, there was a goal to establish 140 MW of CSP;[226] however, no grid-interactive projects were built in that time. Under the Eleventh Plan, India's central government initially targeted 50 MW of solar power, which was to include both CSP and PV. CSP development was to be supported by a preferential tariff offered by MNRE of up to INR 10 (USD 0.20) per kWh for electricity from solar thermal power plants.[227] The system capacity under this scheme was to be between 1 MW and 10 MW per state and 50 MW total. The tariff was to be fixed for a period of 10 years.[228] Now, however, the main push for the development of CSP is expected to come from JNNSM.

4.2.3.1. National Solar Mission for Concentrating Solar Power

With JNNSM, the government will work to create a favorable policy environment for solar thermal manufacturing. The target for Phase 1 (2010–2013) is to establish the technology and set up 500 MW of grid-connected CSP. Since CSP projects have a long gestation period, the entire allocation will be decided upon in financial year 2010–2011. Subsequent phases will attempt to ramp up capacity and achieve economies of scale by aiming for 10,000 MW of grid-connected CSP by 2022.[229]

In February 2010, the CERC announced a feed-in tariff for financial year 2010–2011 of INR 15.3 (USD 0.31) for CSP, which will be revised annually, and declared that PPAs have a validity of 25 years.[230] If total applications for projects (including PV and CSP) in Phase 1 exceed the target of 500 MW, projects will be selected in a process of reverse bidding.[231]

In Phase 1, the minimum project size is 5 MW and the maximum size is 100 MW. A promoter can apply for several projects, but overall, the combined applications may not exceed 100 MW for each promoter. The promoter needs to demonstrate a net worth of INR 30 million (USD 600,000) per MW for the first 20 MW and INR 20 million (USD 400,000) for each further MW.[232]

Projects will be connected to the grid at 33 kV. Plants must use a technology that has been in operation for a period of at least one year or for which financial closure of a commercial plant has already been obtained. At least 30% of the value of the project components (excluding land) has to be obtained from India. Power evacuation will be provided by the state utility.[233]

The promoter will have to pay a non-refundable processing fee of INR 100,000 (USD 2,000), a bid bond [on a graded scale between INR 10,000 (USD 200) and INR 50,000 (USD 1,000) per MW, which rises as the offered tariff lowers], an EMD in the form of a bank guarantee of INR 2 million (USD 40,000) per MW, and later, a performance bank guarantee of INR 3 million (USD 60,000) per MW at the time of signing the PPA.[234]

A project shall achieve financial closure within 180 days from the date of signing a PPA, and all bank guarantees need to be valid for a period of 34 months from the date of signing a PPA.[235]

4.2.3.2 State Policies

In addition to JNNSM, the state of Gujarat has its own CSP policy. In accordance with the Gujarat Solar Power Policy of 2009, the Gujarat Electricity Regulatory Commission set a levelized fixed tariff for solar thermal systems greater than 5 MW to be INR 11 (USD 0.22) per kWh for the first 12 years and INR 4 (USD 0.08) per kWh for the remaining 13 years.[236] The scheme will provide these incentives for solar thermal systems installed and commissioned between 2009 and 2014, for a statewide maximum of 500 MW (of PV and solar thermal combined). There is a control period of two years, which means that eligible projects must be commissioned by March 2012.

Power generated from CSP is exempt from electricity duty. The tariffs have already taken into account the 80% accelerated depreciation tax benefit and would be revised accordingly for those projects that are not availing the tax scheme. In the latter case, projects will be subject to a slightly higher state tariff but can avail the INR 0.50 (USD 0.01) per kWh GBI offered by MNRE.

4.2.4. Technology Development

There is no indigenous capacity for solar thermal power plants, although some manufacturing capacity exists for low-temperature solar collectors. JNNSM announced that incentive packages similar to SIPS (for PV) would be introduced for advanced collectors at low temperatures as well as for CSP collectors at medium and high temperatures.[237]

Under JNNSM, it is envisaged that MNRE will set up demonstration plants for different solar thermal technologies: a 50–100 MW solar-thermal power plant with a storage capacity of 4–6 hours; a 100 MW solar-thermal power plant with parabolic trough technology; a 100–150 MW hybrid plant, which combines solar power with power from coal or biomass; and a 20–50 MW solar-thermal plant with molten salt or steam as a heating medium.[238]

4.2.5. Success Stories from the International Community and Potential Opportunities for India

According to studies done by the German Aerospace Center, CSP on only 0.003 of the world's deserts (i.e., 90,000 km^2) can provide 18,000 TWh/year, which is sufficient for global consumption.[239] However, globally, the use of large-scale CSP for the generation of electricity is still a niche product. Although the technology has been around for a long time, there are currently very few commercially viable plants and capable project developers.

The most famous CSP installation is the set of nine plants totaling 354 MW constructed by Solar Energy Generation Systems in the 1980s in the Mojave Desert in California, United States. Currently in the United States, there are two more CSP plants with a cumulative capacity of 65 MW in operation.[240] In Spain, the Andasol 1 and 2 projects with a cumulative capacity of 100 MW were constructed in 2008 and 2009, respectively, by Solar Millennium.[241]

The DESERTEC Foundation, an initiative of the Club of Rome, is aspiring to tap solar energy in the Sahara Desert to cater to the energy needs of Europe, the Middle East, and Africa.[242] DESERTEC envisions widespread CSP deployment in North Africa and has investigated opportunities for high-voltage direct current transmission lines to carry power produced to southern Europe.

Given the high irradiation data and the endemic power deficit in India, as well as the government's dedication to developing CSP, the Indian market presents very attractive opportunities for strategic investors willing to develop the technology further and make it marketable in India. There is also a significant opportunity in producing equipment like concentrating mirrors or receiving tubes in India in order to bring down the cost of projects.

International companies such as Areva and eSolar, who have advanced thermal technology, are expected to be successful in the Indian market.[243] Siemens, who acquired SOLEL, recently unveiled the latest CSP solar receiver, which will increase the output of parabolic trough solar thermal plants.[244] The company plans to market this to India, among other countries, where there is a potentially high market for CSP.[245]

4.3. Solar Water Heating

SWHs have widespread potential in India including in the residential, commercial and industrial sectors. MNRE has estimated the total SWH potential to be approximately 40 million m^2 of collector area, and as of June 2010, installations had reached just over 3.5 million m^2 of collector area. The targets for 2012 and 2022 are to reach collector area of 5 million m^2 and 20 million m^2, respectively. Government support for SWH deployment is available as capital and interest subsidies. The Indian government is also supporting market development and awareness building through capacity building and outreach activities.

4.3.1. Global Solar Water Heating Capacity

In 2009, the worldwide existing cumulative solar hot water and heating capacity increased by 21% to reach an estimated 180,000 thermal MW (figure excludes unglazed swimming pool heating).[246] In 2008, approximately 70.5% of existing global SWH capacity (149,000 thermal MW) was in China, followed by European Union (EU) (12.3%), Turkey (5.0%), Japan (2.8%), and Israel (1.7%). The Indian share was 1.2%.

4.3.2. Solar Water Heating in India

MNRE estimates that a 100-liter SWH, an appropriate size for domestic use, could save 1,500 kWh of electricity per year. At an energy cost of INR 5 (USD 0.10) per kWh, the savings are INR 7,500 (USD 150). MNRE also estimates that each 1,000 100-liter SWHs deployed can shave 1 MW of peak power demand.[248]

As of 1989, the total collector area of installed SWH in India was approximately 119,000 m^2.[249] Between 1995 and 2008, the annual growth rate in SWH installations was 16.8%.[250] At the end of 2002, India's total installed collector area was 680,000 m^2,[251] and this reached 2.7 million m^2 by the end of 2008.[252] By the end of 2009, the total installed collector area had reached approximately 3.4 million m^2.[253] Approximately 0.5 million m^2 of this value is from sales between April and December 2009.[254]

Table 4-3. SWH Existing Capacity, Top 10 Countries (2008)[247]

Country	SWH Existing Capacity [MWth]	SWH Existing Capacity
China	105,045	70.5%
EU	18,327	12.3%
Turkey	7,450	5.0%
Japan	4,172	2.8%
Israel	2,533	1.7%
Brazil	2,384	1.6%
United States	1,937	1.3%
India	1,788	1.2%
Australia	1,341	0.9%
South Korea	1,043	0.7%
Others	2,980	2.0%
Total	149,000	100%

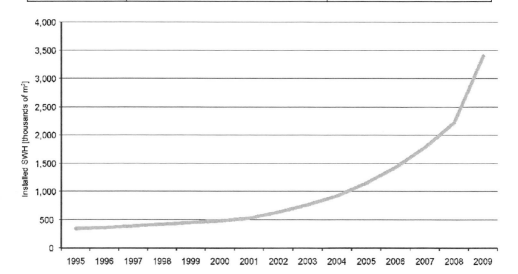

Figure 4-3. Cumulative installation of SWHs in India (1995–2009).

Table 4-4. Estimated Breakdown: Functional SWH Installation Through December 31, 2009[259]

Sector	Percentage Distribution	million m²
Residential	80%	2.108
Hotels	6%	0.158
Hospitals	3%	0.079
Industries	6%	0.158
Other (Railway, Defense, Hostels, Religious Places, Other)	5%	0.132
Total	100%	2.635

At the end of June 2010, India's cumulative SWH installations had reached over 3.5 million m^2 of collector area,[255] less than 10% of the estimated techno-economic potential of approximately 40 million m^2 of SWH collector area.[256] India's current installations work out to approximately 3 m^2 of collector area per 1,000 people. In Israel where 80% of the households use SWH for heating their water, the SWH collector area is approximately 500 m^2 per 1,000 people.[257]

Present installations are concentrated (65%) in two states—Karnataka and Maharashtra—and more than 95% of households in India with SWH are located in urban areas. A 2010 SWH market assessment estimates that 85% of installed SWH are functioning.[258]

As seen in Table 4-4, the SWH installations in India are distributed across the following market-segments: households (residential sector) in both urban and rural areas, commercial and institutional buildings (e.g., hotels, hospitals, hostels, and religious complexes), and industries.

4.3.2.1. Households

In households, hot water is required for bathing, cleaning utensils, washing clothes, cooking, and preparation of cattle feed. The majority of rural households in the country use traditional biomass fuels for most purposes, including heating water. Urban households tend to use liquefied petroleum gas (LPG) and electricity for heating water with the percentage of households using electric water heaters being quite low.

The main use of household SWH is for bathing. The demand of hot water varies significantly across different regions, from 4 months per year in warmer regions to 9 months in colder regions.[260] The highest requirement of hot water in rural households is in the cold region of the country, mostly in the Himalayan states. However, most households with SWH are in non-rural and warmer areas. Only 5% of households with SWH are located in rural areas, and those are mostly in high-income rural households in parts of states including Himachal Pradesh and Kerala. The barriers to increased use of SWH systems in rural areas have been identified as:

- High initial cost of the system.
- Long payback period as the fuel replaced is low-cost biomass fuel.
- Lack of piped water supply.
- Absence of SWH supply chain.
- Difficulty in installing conventional SWH systems on sloping roofs made of metal sheets or thatch.[261]

4.3.2.2. Commercial and Institutional Buildings

In hotels, there is a year-round demand for hot water. The number of hotels with SWH systems installed has increased as a means to reduce the use of expensive petroleum fuels and electricity. Use of SWH for supplying hot water in institutional buildings such as hospitals and hostels has lagged behind the technology adoption observed in the hotel industry.

4.3.2.3. Industries

Hot water and steam are used in a variety of industries including textile, dairy, and pulp and paper. The potential applications of SWH in the industries vary from pre-heating boiler feed water to a temperature of 60°–80° C, heating process hot water up to < 100° C, and canteen applications.[262]

4.3.3. Existing Policies, Measures, and Local Regulations

MNRE has implemented a SWH program to accelerate development and deployment of SWH in India through a combination of financial and promotional incentives. In addition to capital subsidies for SWH systems, this program provides interest subsidies.[263]

MNRE is also implementing a UNDP/UNEP/GEF project titled "Global Solar Water Heating Market Transformation Strengthening Initiative" under UNDP's India Country Program, which is designed to support SWH market development. The overarching objective of the UNDP/UNEP/GEF SWH project is to build on and strengthen the MNRE National Programme and create markets and widespread demand for SWH in different sectors, building awareness in sectors where SWH have application but have not experienced widespread adoption and raising capacity building to increase the quality of the end-use product. This project targets cumulative SWH installations of 10 million m^2 of collector area in India by 2012.[264]

SWHs are an integral part of JNNSM , which targets 20 million m^2 of SWH collectors by 2022. JNNSM targets 3.45 million m^2 of new installations during Phase I to reach 7 million m^2 of installed collector area by 2013, an additional 8 million m^2 installed during Phase 2 ending 2017, and an additional 5 million m^2 during Phase 3 to reach the 20 million m2 target by 2022.

SWH systems have also been incorporated in the new national building code. SWH systems are included in GRIHA rating system[265] for green buildings and energy conservation building codes. An energy labelling scheme similar to the star rating scheme for air conditioners and refrigerators is also planned to promote efficient SWHs.

5. SMALL HYDRO

Section Overview

The estimated potential for small hydro in India of 15,000 MW[266] suggests that it can make a significant contribution to India's power supply, especially in remote areas where alternative supply solutions face many challenges. For these reasons, the further development of small hydro is one of the focal areas of MNRE, who wants to concentrate on reducing the capital costs and enhancing the reliability, plant load factors, and average plant lifetimes. The Indian government aims to develop half of the identified potential in the next 10 years and is supporting small-hydro deployment through capital subsidies and preferential tariffs. As of March 2010, a total of 2,735 MW of grid-connected small hydropower has been installed, contributing about 16.2% to India's total grid interactive renewable power.[267]

Table 5-1. Small Hydro Potential by Indian State, Number of Sites, Installed Capacity, Installed Projects, and Capacity Under Development as of December 31, 2009[272]

State	Number of Prospective Sites	Total Potential Capacity (MW)	Projects Installed (MW)	Projects in Progress (MW)
Andaman & Nicobar Islands	7	7	5	-
Andhra Pradesh	497	560	187	63
Arunachal Pradesh	550	1,329	67	21
Assam	119	239	27	15
Bihar	95	213	55	3
Chhattisgarh	184	993	19	-
Goa	6	7	0	-
Gujarat	292	197	7	5
Haryana	33	110	69	5
Himachal Pradesh	536	2,268	255	185
Jammu & Kashmir	246	1,418	129	6
Jharkhand	103	209	4	35
Karnataka	138	748	588	107
Kerala	245	704	134	24
Madhya Pradesh	299	804	71	20
Maharashtra	255	733	221	67
Manipur	114	109	5	3
Meghalaya	101	230	31	2
Mizoram	75	167	28	9
Nagaland	99	189	29	4
Orissa	222	295	64	4
Punjab	237	393	128	29
Rajasthan	66	57	24	-
Sikkim	91	266	47	5
Tamil Nadu	197	660	90	13
Tripura	13	47	16	-
Uttar Pradesh	251	461	25	-
Uttarakhand	444	1,577	133	234
West Bengal	203	396	98	79
Total	**5,718**	**15,386**	**2,556**	**938**

5.1. Resource/Technological Potential

In India, small hydro plants have a capacity of 25 MW or less and are further subdivided into micro (100 kW or less), mini (between 100 kW and 2 MW), and small (between 2 MW and 25 MW). Large hydropower is not covered in this paper. Small hydropower plants are generally run-of-river, with only small amounts of water stored, if any. These projects are considered environmentally benign, particularly when compared to large hydro plants with storage reservoirs, which can cause habitat destruction and community displacement.[268]

MNRE has estimated the potential for small hydro in India at 15,386 MW for 5,718 prospective plant sites.[269] Of this total potential, over 42% (6,592 MW) is in four northern mountainous states: Himachal Pradesh, Uttarakhand, Jammu and Kashmir, and Arunachal Pradesh. The state-level potential estimates from MNRE are shown in Table 5-1 along with total installed capacity and projects in progress as of December 31, 2009.[270] The current estimate of total small hydro potential increased from the 14,292 MW estimate published in MNRE's 2008–2009 Annual Report,[271] which notes that efforts to identify additional prospective sites are ongoing in both the public as well as the private sector.

5.2. Installed Capacity

The Tenth Plan (2002–2007) targeted the installation of 600 MW of small hydropower. During that period, 520 MW was actually installed. This was significantly less than the 1,438 MW installed during the Ninth Plan (2002–2006).[273] The Eleventh Plan has established a target of 1,400 MW.[274] The goal for the next 10 years is to harness half of the 15 GW of identified potential.[275]

Table 5-2. MNRE Support for New Small Hydro Projects Implemented by the State[279]

Category	Projects 100–1,000 kW	1–25 MW
Special category and NE states	INR 50,000/kW (USD 1,000/kW)	INR 50 million (USD 1 million) for the first MW; INR 5 million (USD 100,000) for each additional MW
Other states	INR 25,000/kW (USD 500/kW)	INR 25 million (USD 500,000) for the first MW; INR 4 million (USD 80,000) for each additional MW

Table 5-1 shows state-level installed capacity of small hydro plants and plants in progress. The state with the greatest amount of installed and in progress small hydro capacity in 2009 was Karnataka, totaling 695 MW, which is 93% of the identified potential in the state. Himachal Pradesh has 440 MW of installed and in-progress small hydro plants, which is less than 20% of the estimated potential in this state. An additional 176.27 MW of small hydro capacity was added between January 1 and March 31, 2010, taking the total for financial year 2009–2010 to 305.27 MW.[276]

5.3. Existing Policies, Measures, and Local Regulations

Until 1989, all hydropower projects were under the administrative control of MoP and the CEA, while the responsibility for execution and maintenance was with the SEBs. Today MoP is involved only in hydropower projects larger than 25 MW, while MNRE is responsible for projects with station capacities of 25 MW or less.[277]

Both the central government and the state governments provide incentives for the construction of small hydropower plants. MNRE proposed a new small hydro scheme in 2009–2010, which provides support in the form of capital subsidies to new plants between 100 kW and 25 MW implemented by the state (see Table 5-2), new plants implemented by the private sector or non-governmental organizations (NGOs) (see Table 5-3), and R&M of existing plants (see Table 5-4).[278] Capital subsidies are also available for watermills and micro hydro projects up to 100 kW. Watermill subsidies range between INR 35,000 (USD 700) and INR 110,000 (USD 2,200). Micro-hydro subsidies range from INR 40,000 (USD 800) to INR 100,000 (USD 2,000). In all cases, special category and northeastern states have a higher subsidy. The amounts are paid in installments, with the schedule of payments varying by the category of project. An additional financial incentive of INR 200,000 (USD 4,000) is available to identify potential sites for projects below 1 MW and an additional INR 500,000 (USD 10,000) for projects between 1–25 MW.

In several states, various additional supporting policies are available for private small hydro projects, including wheeling and banking, buy-back of power, and allowances for third-party sale. Direct subsidies for different project costs are available.[282]

Table 5-3. MNRE Support for New Small Hydro Projects Implemented by the Private Sector and NGOs[280]

Category	Projects 100–1,000 kW	1 MW–25 MW
Special category and NE states	INR 20,000/kW (USD 400/kW)	INR 20 million (USD 400,000) for the first MW; INR 3 million (USD 60,000) for each additional MW
Other states	INR 12,000/kW (USD 240/kW)	INR 12.5 million (USD 250,000) for the first MW; INR 2 million (USD 40,000) for each additional MW

Table 5-4. MNRE Scheme to Renovate and Modernize Existing Projects[281]

Category	Projects 100–1,000 kW	1–25 MW
Special category and NE states	INR 25,000/kW (USD 500/kW)	INR 25 million (USD 500,000) for the first MW; INR 4 million (USD 80,000) for each additional MW
Other states	INR 15,000/kW (USD 300/kW)	INR 15 million (USD 300,000) for the first MW; INR 3.5 million (USD 70,000) for each additional MW

Table 5-5. CERC—Small Hydro Preferential Tariffs for FY 2009–2010[285]

Category	Levelized Tariff INR (USD) per kWh	Net-levelized Tariff after Adjusting for Accelerated Benefit INR (USD) per kWh
Himachal Pradesh, Uttarakhand, NE states (below 5 MW)	3.90 (0.08) (Tariff period 35 years)	3.67 (0.07)
Himachal Pradesh, Uttarakhand, NE states (5–25 MW)	3.35 (0.07) (Tariff period 13 years)	3.14 (0.06)
Other states (below 5 MW)	4.62 (0.09) (Tariff period 35 years)	4.35 (0.09)
Other states (5–25 MW)	4.00 (0.08) (Tariff period 13 years)	3.75 (0.08)

For 2009–2010, CERC has established preferential levelized tariffs for different states and plant capacities (see Table 5-5). Projects below 5 MW often have higher capital and operating costs and cannot take advantage of economies of scale. Though projects in Himachal Pradesh, Uttarakhand, and the northeastern states have a higher capital cost, they also have higher capacity factors due to the hilly terrain and resource availability, so the tariffs offered in these states are less than those offered for projects in other states.[283] The tariff period for small hydro of less than 5 MW has been modified to run for 35 years in order to provide long-term certainty. The tariff period for small hydropower above 5 MW is 13 years, which is the same as for all renewable energy other than solar power.[284] Based on recommendations by CERC, 23 states have declared that they will create attractive investment possibilities for the private sector through incentive schemes. Ten states have already made concrete offers in this direction. Developers also have the option to negotiate special tariffs on a project basis.

In order to be eligible for governmental incentive schemes, a project developer must fulfill different standards of the IEC as well as the International Standards. These standards are:[286]

- For turbines and generators (rotating electrical machines): IEC 60034–1: 1983; IEC 61366–1: 1998; IEC 61116: 1992; IS 4722–2001; IS 12800 (part 3)–1991.
- Field study for hydraulic power of the turbines: IEC 60041–1991.
- Control system of the hydraulic turbines: IEC 60308.
- Transformers: IS 3156–1992, IS 2705–1992, IS 2026–1983.
- Inlet valves of power plants and systems: IS 7326–1902.

5.4. Investment Flows and Industrial Trends

Until the end of the 1990s, the Indian small hydropower market, similar to the entire power market, was dominated by public-sector companies. Most plants even today are operated by the various state power utilities with central government companies playing a smaller role. In recent years, private companies have also started to get involved. Investor interest is increasing due to new incentive systems. As of December 2009, the private sector has accomplished 192 projects with a total capacity of 1,005 MW.[287] This corresponds to 39% of the total installed capacity. In addition, 296 private projects totaling approximately 936 MW are in various stages of implementation.[288] The State of Karnataka leads with 66 private projects holding a capacity of 521 MW.[289]

Most project developers and plant operators are not hydropower specialists but come from the broader energy and infrastructure sector. Examples are HEG Limited and Bharat Petroleum Corporation Limited.

India has a good network of local manufacturers and service providers who can provide complete systems as well as construction and spare parts such as turbines, regulators, and generators.

5.5. Implication in Rural Electrification

Small hydropower is well-suited for rural, remote, and hilly regions, such as the Himalayas, due to the high capacity factors of plants established in these areas.

MNRE has developed a special financial incentives package for on- and off-grid small hydropower in northeast India. As of March 2010, 151 projects, totaling 241.27 MW, have been implemented and another 53 projects, totaling 58.05 MW, were under execution. [290]

During financial year 2009–2010, three new projects totaling 10 MW were approved in Arunachal Pradesh and two projects at 3 MW have been approved for R&M. The ministry has also approved 13 micro hydro projects and 20 watermills in Sikkim.[291]

Arunachal Pradesh Energy Development Agency is executing a program to electrify all border villages in the state through solar, hydro, and micro-hydro systems. This is under the Prime Minister's Special Package worth INR 2.75 billion (USD 55 million) over 3 years. Out of 1,483 villages, 603 will be covered through 62 small hydro projects, 334 through 137 micro-hydro projects, and the remaining 546 through solar PV.[292]

In 2002, Andhra Pradesh set up a governmental agency, Tribal Power Company, for implementing small hydro projects (1–3 MW each) in tribal areas, which are otherwise restricted for development. These projects should be developed in partnership with tribal women, and all profits from the projects are to go to the tribes.[293]

5.6. Technology Development and Transfer

Small hydropower has a capital cost of about INR 50–60 million (USD 1–1.2 million) per MW, which is slightly higher than wind, and a levelized energy cost of about INR 1.50–2.50 (USD 0.03–0.05) per kWh, which is the lowest among renewable energy technologies in India.[294]

Small hydropower equipment has been undergoing steady improvement in efficiency and reliability. This is primarily because of a shift from mechanical to automated electronic control systems and grid integration. Further improvements include remote operating projects and utilization of automatic data collection systems to allow remote monitoring of system performance. The technological trend is to continue to improve reliability and reduce capital cost while increasing efficiency. Logistical and civil construction processes need to be redesigned to reduce installation time. Advancements can also be made in sediment management to reduce silting of equipment.[295]

5.7. Local Case Studies

The Indian industrial company BHEL is a key player in this sector with over 20% of the total installed capacity.[296] BHEL has been involved in the design and manufacture of large hydropower equipment since 1966 (contributing 16,996 MW of capacity to India's total) and with modern renewable technologies since the 1980s.[297] The company has been pioneering the smallest rated bulb and Pelton turbines as well as the highest speed 1,500 RPM Francis turbines.[298]

There are also a number of medium-sized companies in this sector such as Boving Fouress, Escher Wyss Flovel, Jyoti, Steel Industrials Kerala, Kirloskar Bros, HPP Energy, Flovel Mecamidi Energy, Prakruti Hydro Labs, Indusree, Ushvin Hydro System, DRG Jalshakti Eng., Gita Flow Pumps, Pentaflo Hydro Engineers, Everest Energy, Plus Power

System, Standard Electronics Instruments Corporation, and Vinci Aqua Systems. Most offer equipment as well as turnkey solutions and maintenance services. Pentaflo Hydro Engineers is an example of a small hydropower manufacturer based in Delhi. The company manufactures Pelton, Francis, Butterfly Valve, Kaplan, Micro, and Cross Flow turbines for projects from 1 MW to 5 MW.[299] Flovel Mecamidi Energy specializes in complete hydro solutions and has over 20 projects under implementation. The company employs 220 people and had a sales turnover of over INR 900 million (USD 18 million) for financial year 2009–2010.[300]

5.8. Success Stories from the International Community and Potential Opportunities for India

China has made extensive use of small hydro in rural electrification programs, and this sector continues to grow with approximately 33 GW of small-hydro plants (less than 10 MW) added in 2009.[301] Several factors have contributed to the success of small hydro in China, including the low-interest loans provided by the Ministry of Water Resources,[302] government subsidies for capital costs under the Chinese Township Electrification Program, and the effective mobilization of private capital.[303]

The Energy Management Centre in Kerala partnered with UNIDO to establish a regional center for small hydro. The aim of this center is to transfer practices from China's successful small hydro program to the region.[304] The center conducts capacity building events, analysis and design of small-hydro plants, and outreach and coordination regionally and internationally.[305]

International small-hydro companies are not yet very active but are increasingly interested in the Indian market. For example, the project developer Epuron invests in small hydropower in Karnataka. Companies like Alstom, ABB, and Voith Hydro are offering components and engineering services in India.[306]

The Indian market for small hydropower offers good business opportunities to international companies. The market development has been very positive and constant, and the applicability and economic viability of small hydro technology in India has been proven adequately. In order to develop hydropower projects, foreign companies prefer to work with an Indian partner who has a good local network in the relevant region and can take the lead on the purchase of land and the approval processes.

6. BIOENERGY

Section Overview

Historically, traditional biomass has been a major source of household energy in India. Today, the total energy supply in India is composed of approximately 40% non-commercial energy sources such as wood and cow dung.[307] Rural households in India predominantly use wood and cow dung as fuel for cooking and water heating due to lack of electricity.[308] Modern biomass energy is derived from organic material and can be used in a variety of conversion processes to yield power, heat/steam, and fuel. In India, the use is focused on waste materials such as municipal, agricultural, or forest residues. Biomass is generally divided into three categories: biogas, solid biomass, and liquid biofuels.

Biogas has been mostly used for small, rural, and off-grid applications, and the majority of gasifiers in India are providing for individual households. About 4 million family-size biogas plants were installed in India and MNRE estimates that the annual biogas generation potential is about 17,340 million m^3, which could support the installation of up to about 12 million family-size biogas plants. Another 70 projects with aggregated capacity of 91 MW electricity equivalent were installed in India through larger scale biogas facilities use wastewater generated from beverage, meat processing, pulp and paper, food packaging, and other industrial sectors. Another 73 medium biogas plants with aggregated capacity of 461 kW have been installed under Distributed/Grid Power Generation Program of the Ministry. Existing incentives for encouraging biogas production include financial incentives for turnkey operations in rural areas, loans for developing biogas plants in agricultural priority areas, and automatic refinancing offered by the National Bank for Agriculture and Rural Development (NABARD).

Biomass resources in India are used for power generation through three general applications: grid-connected biomass power plants, off-grid distributed biomass power applications, and cogeneration via sugar mill and other industries. The amount of biomass resources in India is estimated about 565 million tonnes per year, including agricultural residues and forest residues. The surplus biomass resources (not used for animal feed, cooking, or other purposes) available for power generation annually is about 189 million tonnes, which could support roughly 25 GW of installed capacity.

India's biofuel strategy is currently focused on using non-food feedstock for the production of biofuels, mainly sugar molasses and non-edible oils, and development and application of advanced conversion technologies are being explored for the near future. The commercial production of biodiesel is very limited, and what is produced is mostly sold for experimental projects and to the unorganized rural sector. The existing biodiesel producers in India are using non-edible oilseeds, non-edible oil waste, animal fat, and used cooking oil as feedstock.

6.1. Biogas

Biogas is obtained via an anaerobic process of digesting organic material such as animal waste, crop residues, and waste from industrial and domestic activities to produce the combustible gas methane. In India, most biogas plants primarily use cattle manure and operate at a household level to meet cooking and lighting needs in rural areas throughout the country. MNRE estimates that with the biogas generation potential, about 12 million family-size biogas plants could be supported. An additional 1,300 MW could be supported from biogas using industrial wastewater (primarily from distilleries and sugar and starch processing plants).

6.1.1. Resource/Technological Potential

Biogas can be combusted directly as a source of heat for cooking, used for space cooling and refrigeration, or used as fuel in gas lamps for lighting. It may also be used to fuel internal combustion engines for production of mechanical work or for electricity. The slurry produced after digestion can be used directly as a valuable fertilizer.

In India, most biogas plants operate at a household level to meet cooking and lighting needs in rural areas. MNRE reports that there are about 4 million family-size biogas plants installed in the country and estimates that the annual biogas generation potential (based on available cattle manure) is about 17,340 million m^3, which could support the installation of up to about 12 million family-size biogas plants. In addition, there are larger plants servicing clusters of houses and whole villages. As of 2006, the total number of community-level biogas plants was 3,902, which is a small number compared to India's 600,000 villages.[309] Given India's large population and high density (Figure 6-1) of cattle, cattle manure has become the primary feedstock for biogas generation at a domestic and community level. With about 28% of the world's total cattle population, India ranks first in the world.

The larger-scale biogas facilities use wastewater generated from beverage, meat processing, pulp and paper, food packaging, and other industries to produce electricity. MNRE reports that 70 projects with an aggregate capacity of 91 MW have been installed under the Energy Recovery from Urban and Industrial Waste programs of the Ministry.[310] The estimated potential from industrial effluent is about 1.3 GW, and potential is expected to reach 2 GW by 2017.[311] MNRE is exploring opportunities for installing anaerobic digesters at sewage treatment plants to generate biogas for electricity production. The agency estimates that the liquid organic waste generated every year by the urban population in India could support over 2.5 GW of installed capacity.

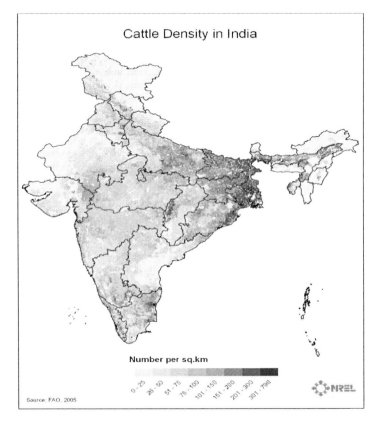

Figure 6-1. Cattle density in India.

Electricity is also produced by biogas plants installed under the Biogas Distributed/Grid Power Generation Program (BGPG) launched by MNRE in 2006. Under this program, the projects are developed at the village level by a community organization, institution, or private entrepreneurs, and the produced electricity is sold to individuals or communities or to the grid on mutually agreed terms. The unit capacity ranges from 3 kW to 250 kW. MNRE reports that, currently, there are 73 projects installed with a total capacity of 461 kW.[312]

6.1.2. Capacity Growth

MNRE reports that 101,529 family-size biogas plants (1–6 m^3 capacity) were installed in India during financial year 2008–2009.[313] This resulted in an estimated savings of about 120,000 tonnes of fuel wood equivalent and generated about 3 million person-days of employment for skilled and unskilled workers in rural areas. Additionally, 51,732 family-size biogas plants were installed between April and December 2009 (out of the planned 150,000 family-size biogas plants for the entire financial year 2009–20 10).

MNRE reported that 10 biogas projects using industrial waste, with a total capacity of approximately 10 MW, were completed during 2008–2009. These included 1.40 MW of installed capacity at a distillery in Tiruchirapally (Tamil Nadu), 3.66 MW using poultry litter in Dupalapudi (East Godavari District, Andhra Pradesh), 0.69 kW at a distillery in Ahmedhnagar (Maharashtra), 1.00 MW using starch waste in Gujarat, and six projects in the states of Maharashtra, Uttarakhand, Madhya Pradesh, Karnataka, and Tamil Nadu with a total capacity of 3.19 MW using liquid waste from the food industry and tanneries. Additionally, eight projects with an aggregate capacity of approximately 14 MW were completed in 2009. These included five projects totalling approximately 12 MW installed capacity at distilleries, two 1 MW projects using waste from starch factories, and one small project generating biogas from yeast waste.

Under BGPG, 65 new projects with a total installed capacity of 0.353 MW were initiated in the states of Andhra Pradesh, Chhattisgarh, Kerala, Maharashtra, and Uttarakhand during 2008–2009. Approximately 14 additional projects were initiated during this time in Tamil Nadu and another 11 projects in Kerala. During 2009–2010, one new project was initiated in Haryana and several project proposals were received from the states of Andhra Pradesh, Goa, Maharashtra, Tamil Nadu, and by Khadi and Village Industries Commission.

6.1.3. Existing Policies, Measures, and Local Regulations

Under the Eleventh Plan, the Government of India established a goal of deploying 2 million family-size biogas plants.[314]

Currently, incentives for biogas projects include:[315]

- Financial incentives for turnkey operations in rural areas.
- Loans for developing biogas plants in agricultural priority areas.
- Automatic refinancing offered by NABARD.

The National Biogas and Manure Management Program (NBMMP), implemented in 1981, continues to support the deployment of biogas plants utilizing manure and other organic wastes.[316] The financial support ranges from INR 2,100 (USD 42) to INR 11,700 (USD 234) per plant. This program focuses on family-size installations to meet the cooking and energy needs of households and the manufacturing of fertilizers in rural areas. Under the NBMMP, 12 biogas training centers have been established throughout India to develop local technical capacity in this field.

The BGPG focuses mainly on the promotion of small capacity biogas-based power generation. In 2008, MNRE established a demonstration program, the Demonstration of Integrated Technology Package on Biogas Generation Purification and Bottling, to determine the feasibility of integrating "generation, purification/enhancement, bottling and piped distribution of biogas" at medium-sized facilities.[317] Under the first phase of this program, MNRE provided financial support up to 50% of the project costs.

MNRE provides incentives for the recovery of energy from MSW and other urban waste streams under the Programme of Energy Recovery.[318] In 2010, five MSW pilot projects are eligible to receive INR 20 million (USD 400,000) per MW installed, up to 20% of the project costs, or INR 100 million (USD 2 million) total, whichever is lower. MNRE also provides 40% of the total project costs with a maximum incentive of INR 20 million (USD 400,000) per MW for the generation of electricity from biogas at sewage treatment plants. For power generation from biomethanation-based projects, MNRE will provide up to 30% of the project costs with a maximum incentive of INR 30 million (USD 600,000) per MW. Both private and public sector projects are eligible for these waste-to-energy incentives.

For the industrial sector, MNRE provides support for the recovery of energy from wastes.[319] Under the Programme on Energy from Industrial Wastes, MNRE provides financial assistance for resource assessment, R&D, technology improvements, performance evaluation, and other associated components of waste-to-energy projects. Industrial sector waste-toenergy projects are eligible to receive incentives between INR 5–10 million (USD 100,000– 200,000) per MW. These incentives are available for both private and public sector projects.

6.1.3.1. Local Policies

The State of Haryana announced a program promoting the installation of 50,000 family-size biogas plants. The program is designed to take advantage of CDM funding by having a single financial institution undertake the project as a whole instead of focusing on individual project installations by separate entities.[320]

6.1.4. Investment Flows and Industrial Trends

The demand for and number of small technical and thermal gasification plants is increasing in India. The wide ranges of manufacturers in this market are steadily improving their technology and products, which has lead to an establishment of Indian companies in the global market.[321] Thus, foreign companies face strong competition from already established Indian companies in the Indian market who have knowledge of local conditions and requirements.

For larger scale projects on the MW-scale level, biogas plants have not yet been widely commercialized. Given the significant potential for biogas from various sources of agricultural or industrial waste, companies able to develop reproducible and sustainable business models for larger projects have a great market opportunity in India. A further step of value creation can be added by upgrading biogas to natural gas equivalent and feeding it into the grid. This technology is not applied yet in India.

Some international companies are now venturing into this field. In the scope of a public-private partnership, EnviTec, a German biogas specialist, and its Indian partner MPPL, with the support of GTZ, are planning a pilot project in Punjab. The project is to have a capacity of 25 MW, later expanding to 750 MW.[322]

6.1.5. Technology Development and Transfer

During 2008–2009, MNRE established a new initiative to demonstrate an integrated technology for biogas plants at a medium scale (200–1,000 m^3/day). This technology integrates biogas generation, purification/enrichment, bottling, and piped distribution. During 2009–2010, eight integrated biogas plants at a medium scale, with an aggregate capacity of 5,700 m^3/day, were initiated in the states of Bihar, Chhattisgarh, Gujarat, Haryana, Karnataka, Maharashtra, and Punjab.[323]

6.1.6. Local Case Studies

There are some interesting cases of companies who develop biogas solutions on a mass scale. One example is the Indian company SKG Sangha, who develops small biogas plants for domestic use in Karnataka. Animal and household wastes are used as feedstock, thereby replacing firewood with sustainable and less-polluting sources and improving waste disposal.

By 2009, over 80,000 plants had been installed, each saving around 4 tons of CO2 per year. The plants were built and maintained by 2,000 employees.[324]

Another successful case study comes from Kerala, where 16,000 plants were installed by the Indian company BIOTECH by 2009. Home, market, and municipal waste are used as feedstock, and the plants vary in size. The installation costs can be partly reduced through government subsidies, and an average family is able to finance a plant within 3 years.[325]

6.2. Solid Biomass

Solid biomass includes agricultural and forest residues as well as organic household and industrial wastes for direct combustion or gasification to provide electricity or combined electricity and heat (cogeneration). MNRE estimates that the surplus biomass resources potentially available for power generation could support roughly 25 GW of installed capacity.[326] Cogeneration plants are mainly found in the sugar industry where the heat can be used for the production process and the power surplus is typically sold to the grid. MNRE estimates that there is a potential of about 15 GW using cogeneration in various industries including sugar mills, breweries, textile mills, distilleries, fertilizer plants, pulp and paper mills, and rice mills.

6.2.1. Resource/Technological Potential

Biomass resources in India are used for power generation in three general applications:

- Grid-connected biomass power plants (using combustion and gasification conversion technologies).
- Off-grid/distributed biomass power applications (using primarily gasification conversion technology).

- Cogeneration (simultaneous production of both heat energy and electricity from one energy source).
  - Bagasse cogeneration in sugar mills.
  - Non-bagasse cogeneration in other industries.

6.2.1.1. Power Generation

MNRE estimates the amount of biomass resources in India at about 565 million tonnes per year, including agricultural residues (resulting from crop harvesting and processing) and forest residues (resulting from logging and wood processing). The agricultural residues, which provide most of the biomass resources in the country, include rice husk, rice straw, bagasse, sugar cane tops and leaves, groundnut shells, cotton stalks, and mustard stalks. Some of the agricultural and forest residues are already in use as animal feed and fuel for domestic cooking, among other purposes. MNRE estimates that the amount of surplus biomass resources available for power generation annually is about 189 million tonnes, which could support roughly 25 GW of installed capacity.[327] As illustrated in Figure 6-2, states with the highest potential include Maharashtra, Madhya Pradesh, Punjab, Gujarat, and Uttar Pradesh. Total biomass power installed capacity in India is about 829 MW.

6.2.1.2. Cogeneration

Cogeneration plants provide both heat energy, used in the mill, and electricity, which is typically sold to the grid. MNRE estimates that about 15 GW of electricity generating capacity could be achieved through adding cogeneration capabilities in various industries including sugar mills, breweries, textile mills, distilleries, fertilizer plants, pulp and paper mills, and rice mills. Alternatively, gasifier systems that convert biomass resources into a combustible gas for the generation of electricity and heat could be installed at these plants to meet their captive energy requirements.

India is the world's second largest sugar producer with production companies concentrated in the sugarcane-growing states of Uttar Pradesh, Maharashtra, Gujarat, Tamil Nadu, Karnataka, and Andhra Pradesh. Uttar Pradesh accounts for 31% of the total sugar production in the nation and Maharashtra's contribution is close to 28%. The Indian government encourages sugar mills to invest in cogeneration plants using bagasse, the fibrous residue remaining after sugarcane stalks are crushed to extract their juice. MNRE reports that there are about 550 sugar mills in the country with a potential for 5 GW installed capacity.

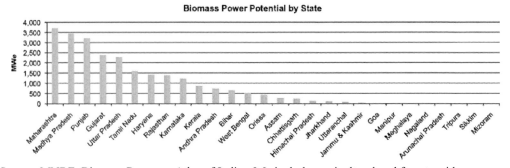

Source: MNRE. Biomass Resource Atlas of India v2.0; includes agricultural and forest residues.

Figure 6-2. Biomass potential by state.

6.2.2. Capacity Growth

In 2008, the grid-connected capacity additions for biomass power and bagasse cogeneration were 97 MW and 248 MW, respectively.[328] Between April and December 2009, biomass power and bagasse cogeneration contributed 384 MW of capacity additions bringing the total biomass power and cogeneration capacity to 2,136 MW.[329] Biomass power projects contributed 125 MW and bagasse cogeneration contributed 259 MW during this period. The total installed grid-connected biomass power and bagasse cogeneration capacity was 2,322 MW at the end of June 2010.[330]

MNRE reports that in the last 4 years, 63 non-bagasse cogeneration and 153 biomass gasifiers were brought online with total capacity of 211 MW and 45 MWe, respectively, to provide power to industrial facilities.[331] The cumulative installed capacity as of June 30, 2010, was 238 MW of biomass power and non-bagasse cogeneration and 125 MW of biomass gasifiers.[332]

6.2.3. Existing Policies, Measures, and Local Regulations

During the Eleventh Plan, the central government plans to add 2,100 MW of grid-connected biopower, representing an increase of about 37% in total capacity installed compared to 2007.[333]

Incentives for biomass power projects include capital subsidies, tax benefits, and preferential tariffs.

6.2.3.1. Tariff for Electricity Generated from Biomass

For 2009–2010, CERC has established a levelized preferential tariff for biomass energy projects ranging from INR 3.93/kWh (USD 0.079/kWh) to INR 5.52/kWh (USD 0.11/kWh).[334] The levelized tariff decreases to INR 3.35 (USD 0.064) to INR 4.62/kWh (USD 0.092/kWh) in 2010–2011 (including available depreciation benefits, the net levelized tariff is INR 3.16 [USD 0.063] to INR 4.43/kWh [USD 0.089/kWh]).

Biomass projects may also qualify for 80% depreciation on specific equipment such as fluidized bed boilers and high efficiency boilers, a 10-year income tax holiday, and central excise duty exemptions.[335]

6.2.3.2. Biomass Energy and Cogeneration Incentives

Under the MNRE Biomass Energy and Cogeneration in Industries Programme, projects that utilize currently untapped resources to provide captive thermal energy and electricity can receive INR 200,000 (USD 4,000) to INR 1.5 million (USD 30,000) per 100 kW project for biomass gasifier systems and up to INR 2 million (USD 40,000) per MW for non-bagasse biomass cogeneration projects. The energy captured in these projects must be used onsite. Furthermore, financial incentives such as "accelerated depreciation, concessional custom duty, excise duty exemption, tax exemption, income tax exemption on projects for power generation for 10 years... are available to biomass power projects."[336]

6.2.3.3. Local Policies

Local support for biomass power generation varies by state.[337] Preferential tariffs for biomass projects are available in several states including Andhra Pradesh, Haryana, Madhya Pradesh, Maharashtra, Punjab, Rajasthan, Tamil Nadu, Bihar, Karnataka, West Bengal,

Chhattisgarh, Gujarat, Uttarakhand, Kerala, and Uttar Pradesh ranging from INR 3.35/kWh (USD 0.067/kWh) in Madhya Pradesh to INR 4.62/kWh (USD 0.092/kWh) in Haryana.[338] Some states provide an exemption from the general sales tax, and some provide capital subsidies up to INR 2.5 million (USD 50,000) per MW.[339]

6.2.4. Investment Flows and Industrial Trends

According to MNRE, combustion of biomass and cogeneration has been used in the sugar industry for over 20 years, mainly for waste disposal. Since the primary aim was not to produce energy, the efficiency of these plants is low. However, the possibility to generate an additional benefit through feeding the energy surplus in the grid leads to new dynamics and new opportunities. Large companies like Triveni or Simbhaoli Sugars are changing from being solely sugar producers to also providers of energy technology and power.

Companies from other industries are realizing the business opportunities in biomass for large- scale generation of electricity as well. A2Z Maintenance & Engineering Services, an Indian EPC company, is planning to set up a 500 MW power plant that uses municipal waste.[340]

As in the case of biogas, a proven market-entry strategy for foreign companies is in partnerships with local companies. The U.S.-based Indus Terra, for example, cooperates with Indian Emergent Ventures in the form of a special purpose vehicle: Green Indus Bio Energy Private Limited. The partnership is implementing a power project with a capacity of 5.6 MW, which will use poultry waste to generate electricity for the state power grid and slurry for fertilizer. A cluster of 150 poultry farms in Haryana, which can sell their waste to the plant, supports the initiative. Technology used will be designed in the UK, but the machines will be assembled in India.[341]

6.2.5. Local Case Studies

To tackle the problem of power cuts in Bihar, the local energy company, Saran Renewable Energy, has built a power plant that runs on gasified biomass from a local type of bush that farmers can easily and profitably grow. The generator connects to transmission lines to supply small businesses with electricity, hence providing energy security. The prices for users are higher than the electricity from the grid, but the costs are still much less then for generating power by using diesel.[342]

Nishant Bioenergy developed a combined stove fuelled by crop waste and used in schools to cook meals for pupils. Local farmers that produce the waste sell it to local plants that compress it into briquettes for fuel use. Farmers are able to increase their income by an estimated 10%, and the schools save on fuel costs. Nishant provides credit for the purchase of the stove, installs the stove, provides a 3-day training, and offers an annual maintenance contract.[343]

Winrock International India started a remote village electrification initiative, which aims to promote alternative bioenergy in Chattisgarh. It uses Jatropha oil and is designed to be a replicable model for helping communities to be energy independent. Implemented in 2007, the initiative has served 107 households, or 553 people, with 3 hours per day of electricity generated for each household and 3.5 hours for streetlights.[344]

6.3. Liquid Biofuels

Liquid biofuels, namely ethanol and biodiesel, are used to substitute petroleum-derived transportation fuels. India's biofuel strategy is focused on using non-food sources for the production of biofuels: sugar molasses and non-edible oilseeds as well as second generation biofuels in the near future. Advanced conversion technologies for ethanol are under development, which will allow it to be made from forest and agricultural residues. Using one-third of the surplus, biomass could yield about 19 billion liters of ethanol, which could displace the country's entire gasoline consumption once techno-economically viable. Advanced feedstock for biodiesel production includes microalgae, which is currently under research and has a very promising potential in India—it can provide double the yield of the highest producing crop (oil palm) per land unit.

6.3.1. Resource/Technological Potential

6.3.1.1. Ethanol

In India, ethanol is produced by the fermentation of molasses, a by-product of the sugar industry. There are about 320 distilleries producing ethanol for industrial, beverage, and other purposes with a total production capacity of about 3.5 billion liters per year. More than 115 distilleries have modified their facilities to produce fuel grade ethanol with a total capacity of 1.5 billion liters per year.[345] This production capacity is sufficient to meet the estimated ethanol demand of 800 million liters for a 5% blend with gasoline.[346] However, for a 10% blend, the fuel grade ethanol production capacity from molasses needs to be expanded or the industry could consider direct production of ethanol from sugarcane juice, which offers higher productivity but will require additional investments for technological modifications.[347] The direct production of ethanol from sugarcane juice requires an increase in feedstock production, which in India means developing higher yields per land unit rather than an increase in area planted with sugarcane because this crop is very water intensive and the country's irrigation water supplies are increasingly limited.[348]

Efforts to produce ethanol from other feedstock such as sweet sorghum, sugar beet, and sweet potatoes are at an experimental stage in India.[349] Additionally, various public and private institutions in India are conducting research in the area of cellulosic ethanol, which uses feedstock such as agricultural and forest residues. As mentioned previously, India has a large amount of this biomass material and is therefore a promising potential for the production of cellulosic ethanol. Using one-third of the 189 million tonnes surplus of biomass could yield approximately 19 billion liters of ethanol (assuming one tonne of lignocellulosic biomass = 300 liters of ethanol), which could displace the country's entire gasoline consumption (approximately 14 billion liters in 2008–2009).[350]

6.3.1.2. Biodiesel

India's commercial production of biodiesel is very small, and what is produced is mostly sold for experimental projects and to the unorganized rural sector. Although India's biodiesel processing capacity is currently estimated at 200,000 tonnes per year, the majority of biodiesel units are not operational during most of the year. The existing biodiesel producers in

India are using non-edible oilseeds, non-edible oil waste, animal fat, and used cooking oil as feedstock.[351]

Biodiesel production efforts in India are focused on using non-edible oils since the demand for edible oils exceeds the domestic supply. It is estimated that the potential availability of non-edible oils in India amounts to about 1 million tonnes per year. The most abundant resources are sal oil (180,000 tonnes), mahua oil (180,000 tonnes), neem oil (100,000 tonnes), and Karanja oil (55,000 tonnes). However, based on extensive research carried out by various institutions in the country, the government identified *Jatropha Curcas* oilseed as the major feedstock for biodiesel in India.[352]

Since early 2001, the Ministry of Rural Development and several state governments have carried out programs to encourage large-scale planting of Jatropha on wastelands. States with the largest potential include Madhya Pradesh/Chhattisgarh, Rajasthan, Maharashtra, Andhra Pradesh, and Gujarat (Figure 6-3). The Indian government's Planning Commission set an ambitious target of 11 million ha of Jatropha to be planted by 2012 in order to generate sufficient biodiesel to blend at 20% with petro-diesel. However, the total Jatropha plantation area in the country is currently estimated at approximately 450,000 ha, of which about 60%–70% are new plantations (1–3 years old) and not yet into full production. The new Jatropha plantations are expected to come into maturity in the next 3–4 years.[353]

There is no commercial production of biodiesel from other resources such as microalgae, another promising feedstock, except for some experimental trials by research organizations.

6.3.2. Capacity Growth

6.3.2.1. Ethanol

The ethanol supply during 2008–2009 was severely impacted by the short supply of sugar molasses and continued strong demand for alcohol from other competing industries. Consequently, the ethanol available for blending with petrol was about 100 million liters, considerably short of the 600 million liter target. This trend has continued through 2010, and the supply of ethanol for fuel was even lower at 50 million liters.[355]

6.3.2.2. Biodiesel

According to the Biodiesel Association of India, biodiesel production in the country during 2009–2010 was approximately 80,000 tonnes, sold entirely to commercial establishments using diesel generators.

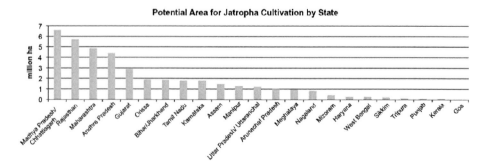

Figure 6-3. Potential area for Jatropha cultivation by state[354].

6.3.3. Existing Policies, Measures, and Local Regulations

In 2003, India's Ministry of Petroleum and Natural Gas (MPNG) mandated 5% ethanol blending with petrol in nine states and four union territories under Phase 1 of the ethanol blended petrol (EBP) program.[356] Under Phase 2, this mandate was expanded to include 20 states and eight union territories in 2006. Due to supply shortages and extended negotiations between ethanol suppliers and state governments over state taxes and other fees, ethanol costs were too high to be commercially viable. As only domestically produced ethanol qualifies for the EBP program, and there continues to be a shortage of ethanol in India, the government has postponed implementation of Phase 3, which would raise the blended ethanol requirement to 10% in the participating states and require the remaining states to have at least a 5% ethanol blend ratio. The central government does not currently provide any direct financial or tax incentives for the production of ethanol but does offer subsidized loans for some sugar mills to set up ethanol production systems. The government does fund ethanol R&D programs.

The Eleventh Plan set a goal to achieve at least a 5% blend of biofuels for petrol by the middle of the plan period and to potentially increase the blend to 10% by the end of the Eleventh Plan.[357] The plan also suggested that a 5% blend of biodiesel might be possible by the end of the Eleventh Plan depending on availability. Proposed funding for biofuels R&D from the National Energy Fund is INR 2 billion (USD 40 million). The Planning Commission established the goal for 11.2–13.4 million ha to be cultivated with Jatropha to generate sufficient biodiesel to achieve a 20% blend with petro-diesel.[358] The government's initial demonstration phase brought 400,000 ha under production, and the plan is for production to be expanded in the coming years to meet the target by 2012. The MPNG established a program with 20 diesel procurement centers to purchase enough biodiesel to attain a 5% blend ratio with high-speed diesel. However, the high cost to produce biodiesel has resulted in no biodiesel sales at these centers. Like for ethanol, the central government does not provide direct financial incentives for biodiesel production. However, biodiesel is exempted from the central excise tax and some state governments provide similar exemptions. Further, the central government and multiple state governments provide financial incentives to promote the cultivation of Jatropha and other non-edible oilseeds that may be used to produce biodiesel.

MNRE announced a National Policy on Biofuels in December 2009, which details targets, strategies and approaches, incentive levels, and policies for promoting the development and use of biofuels.[359,360]

- Target 20% blending of biofuels with petrol and diesel by 2017.
- Promote domestic biofuel production.
- Limit cultivation of biofuels to wastelands.
- Establish minimum support prices and minimum purchase prices for biofuels.
- Establish a National Biofuel Coordination Committee and a Biofuel Steering Committee to provide high-level coordination and policy guidance on the topic of biofuels and to oversee implementation of the program.

To support the national mission on Jatropha biodiesel, many states have announced extensive Jatropha cultivation programs such as Andhra Pradesh, Rajasthan, Maharashtra, Gujarat, Tamil Nadu, Chhattisgarh, and Haryana.[361]

6.3.4. Investment Flows and Industrial Trends

More than 20 companies are producing biodiesel in India with a combined capacity of 1 MT per year. A 2009 directive from MPNG banned the use of biodiesel for transportation, so biodiesel producers have been forced to look for new markets for their product as a fuel for electricity generation. The West Bengal Pollution Control Board issued a directive that all telecom towers should use a minimum of 30% biodiesel to power their operation. At least one company has made supply arrangements with one telecom tower operator and is in discussion with more.[362]

6.3.5. Technology Development and Transfer

In India, efforts to develop advanced ethanol and biodiesel technologies are underway by various public and private research institutions, such as Praj Industries Ltd. (cellulosic ethanol and biodiesel), Indian Oil Corporation (cellulosic ethanol and algae), and the Department of Biotechnology (advanced feedstock development). These institutions work with research entities abroad to develop and transfer technologies.

6.3.6. Local Case Studies

Indian Railways is the largest consumer of diesel in the country with an annual expenditure on diesel fuel of INR 65 billion (USD 1.3 billion). Due to the high potential of cutting costs by using alternative sources, Indian Railways plans to use biodiesel (from Jatropha and other non-edible sources) for powering their locomotives. It will set up an integrated biodiesel plant in Chennai with a production capacity of 30 tonnes per day to gradually reduce dependence on high speed diesel, which is what is used now. Indian Railways would like to start with a 5% biodiesel blend and increase this up to 20% in coming years. The project is to be executed under a public-private partnership model and completed over 2 years. While the main goal is to reduce costs, other priorities include reducing GHG emissions and qualifying for carbon credits.[363]

6.3.7. Success Stories from the International Community and Potential Opportunities for India

Betim, Brazil (population 440,000), is one of Brazil's first "model communities" for renewable energy, as part of a six-cities network that also includes Belo Horizonte, Porto Alegre, Salvadore, São Paulo, and Volta Redonda. Betim has established a number of policies to promote biofuels use in transportation. The city mandates biofuels in public buses and taxis and also gives preference to flex-fuel vehicles for municipal vehicle fleet purchases. The city is also facilitating the addition of SWH systems to a low-income housing project being built under a national program. A number of demonstration projects on municipal buildings have been carried out. Betim has also established a Renewable Energy Reference Center that raises public awareness; provides information; brings together diverse stakeholders from local, state, and national levels; conducts training and workshops; and conducts outreach to other local communities in Brazil to share Betim's experience."[364]

7. DECENTRALIZED ENERGY

Section Overview

Renewable technologies are ideally suited to distributed applications, and they have substantial potential to provide a reliable and secure energy supply as an alternative to grid extension or as a supplement to grid-provided power. According to IEA, in 2008, over 400 million people in India, including 47.5% of those living in India's rural areas, still had no access to electricity. Because of the remoteness of much of India's un-electrified population, renewable energy can offer an economically viable means of providing connections to these groups. The Government of India has initiated several programs, policies, and acts to improve rural livelihood with the help of renewable energy. They support electricity provision programs and organizations aiming to meet the full power requirements of a village including motive power and cooking energy. These programs have resulted in electrification of more than 90,000 villages and hamlets and free electricity connections have been provided to more than 12 million below poverty line (BPL) households. Renewable technologies also present opportunities for industries looking to establish captive power plants at their facilities to supplement grid power. Currently, renewable technologies represent only 1.2% of the total captive power installed capacity in India.

7.1. Renewable Energy—Implication in Rural Electrification

The lack of available modern energy services is a primary limitation to development, and providing access to electricity in a country of over one billion people, of which the majority lives in rural areas, is a gigantic task. Currently, many potential customers live in areas that are either physically inaccessible for a centralized power grid connection or where a connection is economically unviable, so creating decentralized energy supply solutions will be essential for stimulating development in these areas. In India, locally available, renewable sources such as solar energy, wind, biomass, or hydro energy are playing an increasingly prominent role in extending access to energy.

Sustainable electricity supply is mainly necessary for lighting, cooking, and comfort but also for watering agricultural fields. Growing demand due to the increase in population and more energy-intensive lifestyles has diminished the availability of dominantly used fuels in rural areas, such as wood, coal, and kerosene, which further underscores the need to shift to more sustainable energy supplies. Government initiatives to utilize locally available energy resources need private sector support and collaboration between various stakeholders to achieve the goals of energy security and energy sufficiency in rural areas of India.

At present, access to and availability of reliable and assured energy in India's rural areas is inadequate. According to IEA, in 2008, over 400 million people in India, including 47.5% of those living in India's rural areas, still had no access to electricity.[365] Even those with electricity connections may periodically be unable to use it due to shortages or poor quality of supply resulting from inadequate infrastructure or inability to pay.

Households without access to electricity, whether in the electrified or un-electrified villages, continue to rely mostly on inefficient traditional biomass or kerosene. Fuel-wood, chips, and dung cakes contribute approximately 30% of the commercial energy consumed.[366]

To meet cooking energy needs, traditional biomass, heavily subsidized LPG, and kerosene are used by approximately 90.1%, 8.6%, and 1.3% of rural households, respectively. [367] Additionally, approximately 40%–60% of rural households use kerosene for lighting.[368]

Some of the renewable energy technologies that are used in villages and rural areas as decentralized systems are:

- Family-size biogas plants.
- Solar street lighting systems.
- Solar lanterns and solar home lighting systems.
- Solar water heating.
- Solar cookers.
- Standalone solar power generators.
- Wind pumps.
- Micro-hydro power systems.

7.1.1. Existing Policies and Relevant Programs for Rural Electrification

As per the Electricity (Supply) Act 1948,[369] in force until 2003, rural electrification was the responsibility of the state governments and SEBs. Therefore, non-conventional and renewable energy sources for these purposes were also promoted by the SEBs.

The Electricity Act 2003 made a specific provision of introducing national policies on standalone non-conventional energy systems in rural areas and electrification with local distribution to rural areas, and the pace of development of renewable energy sources has since increased substantially since this time.

The National Electricity Policy (February 2005) recognizes electricity as an essential requirement for all facets of life and as a basic human need. It also states that electricity supply at a reasonable rate to rural India is essential for India's overall development.

Three nodal ministries in the Government of India are focusing on the development of the rural economy by tapping local resources and providing subsidies and grants to energize growth and economic development.

- MoP: Promotes rural electrification.
- MNRE: Promotes use of renewable energy.
- Ministry of Rural Development: Contributes to the living conditions, business, and economy in rural areas.

In October 1997, the Government of India defined "A village will be deemed to be electrified if electricity is used in the inhabited locality within the revenue boundary of the village for any purpose whatsoever."[370] The Indian government revised this in February 2004 so that villages must meet each of the following criteria before it is considered electrified:

"a) Basic infrastructure such as Distribution Transformer and Distribution Lines are provided in the inhabited locality as well as a minimum of one Dalit Basti/hamlet where it exists; and

b) Electricity is provided to public places like Schools, Panchayat Office, Health Centers, Dispensaries, Community Centers etc.; and

c) The number of households electrified are at least 10% of the total number of households in the village."[371]

This changed the electrification status of many villages who were previously deemed electrified.

The Government of India has initiated several programs, policies, and acts that focus on the development of rural energy, economy, and electrification to improve rural livelihood with the help of renewable energy.

7.1.1.1. Rajiv Gandhi Grameen Vidyutikaran Yojana —2005

RGGVY is supported by MoP through the Rural Electrification Corporation. The targets of this program are to electrify 125,000 villages and to provide free electricity connections to 23.4 million BPL households. [372] The program provides a 90% capital subsidy with the remainder of funds provided as a loan through the Rural Electrification Corporation.[373] According to the International Energy Agency, the RGGVY is also designed to strengthen the existing network in 462,000 electrified villages until 2010[374] by providing:

- Rural electricity distribution backbone for a cluster of villages by providing 33/11 kV or 66/11 kV substations of adequate capacity.
- Creation of village electrification infrastructure in every village including adequate distribution transformers.
- Decentralized distributed generation (DDG) and supply to provide electricity to villages from conventional or renewable sources on a standalone basis.

As RGGVY is focused on the provision of electricity and not the generation source, renewable energy is not necessarily promoted under this program; however, the DDG provision includes the use of renewable technologies.[375] As of September 30, 2010, 84,618 villages have been electrified and over 12 million BPL households have received electricity connections.[376]

7.1.1.2. Solar Mission

JNNSM is supported by MNRE and MoP (see Chapter 4 for an overview of JNNSM). In addition to JNNSM's goals of 20 GW of grid-tied solar by 2022, it aims to reach 2 GW of off-grid solar installed capacity in the same time period. Under JNNSM, the first 1,000 MW of off-grid solar is to be installed by 2017.[377] The central government is providing capital subsidies and low-interest loans for off-grid PV projects up to 100 kW for an individual site and up to 250 kW for a solar mini-grid. Subsidies are also available for off-grid solar thermal applications.[378] As of August 31, 2010, almost 20 MW of projects had been sanctioned by MNRE, more than half of which are in the state of Rajasthan.[379]

7.1.1.3. Village Energy Security Programme

The Village Energy Security Programme (VESP) is supported by MNRE and is designed to support all energy needs of villagers including domestic, commercial, agricultural, industrial, and motive power through test projects using locally available biomass. This program targets villages between 25–200 households and aims to support micro-enterprise development by providing energy for productive uses. To this end, it provides a 90% capital subsidy; operation and maintenance expenses are to be met through usage charges, and all remaining expenses are to be secured by the community, implementing agency, or state nodal agency.[380] At the end of 2009, 80 test projects had been approved under VESP, and 54 of these had been commissioned. MNRE plans to complete the already approved test projects but currently has no plans to expand the program further.[381]

7.1.1.4. Remote Village Electrification

Remote Village Electrification (RVE) is supported by MNRE and aims to provide electricity to hamlets and villages with more than 300 inhabitants not covered under RGGVY and more than 3 km from the nearest point of grid access. The technology deployed under RVE is chosen by the state nodal agency after an assessment of renewable resource availability, and the Indian government provides a subsidy for installation up to 90% of actual costs.[382] As of June 30, 2010, RVE activities had been completed for 6,867 hamlets and villages.[383]

7.1.1.5 Programs of National Bank for Agriculture and Rural Development

NABARD is set up as a development bank with the support of the Ministry of Finance with a mandate for facilitating credit flow for promotion and development of agriculture, small-scale industries, cottage and village industries, handicrafts, and other rural crafts (see Chapter 9).

7.1.2. Business Models for Rural Electrification

The Electricity Act 2003 gave private investors access to all power sector operations. As a consequence, state governments encouraged the private sector to invest in rural electrification projects. However, at present, the private sector is not independently active in village energy- based commercial activities.

Many business models for rural economy development through renewable energy sources operate with the support of government programs. Some of the government programs succeeded, yet several programs did not fulfill the desired objective as revenue generation from these projects did not yield any tangible benefits.

Currently, a few of the business models are guiding progress towards desired objectives, such as SELCO (solar), NTPC (distributed generation), Appropriate Rural Technology Institute (ARTI) (biomass stoves), and ANKUR (scientific biomass). The following describes some of these business models.

7.1.2.1. Biomass Gasifiers

MNRE implements a program called "Biomass gasifier-based distributed/off-grid power programme for rural areas."[384] This program to deploy biomass gasifier systems for meeting unmet electricity demand in villages is implemented through state nodal agencies[385] with the involvement of either energy service companies, cooperatives or Panchayats, NGOs, manufacturers or entrepreneurs, and central finance assistance (CFA). State nodal agencies are responsible for initiating, promoting, supporting, and coordinating all activities concerning new and renewable energy within each state. Some components, which are funded by the CFA, focus on human resource development and training (operation and maintenance;

management) and support for gasifier manufacturers and suppliers to establish service centers in areas near the location that the biomass gasifiers are installed.[386]

Sustainable funding models are also developed in the villages, although the capital costs for the biomass gasifiers are funded through Government of India grants. The payable electricity charge is collected from each household, and the operation and maintenance costs are met by the Village Energy Committee or the entrepreneur. Figure 7-1 depicts material and financial flows for a village-level biomass gasifier.

7.1.2.2. Biomass Stoves

ARTI developed a project to disseminate improved cook stoves and improved fuels while also building up a sustainable network of manufacturers and marketers of the technology. At the end of the pilot stage, 120 rural businesses across Maharashtra had sold improved stoves to approximately 75,000 households. ARTI plans to scale up the program to reach 1.5 million households in Maharashtra and 50,000 in Gujarat with engagement of self-help groups, rural entrepreneurs, and NGOs.[387]

7.1.2.3. SELCO Business Model

SELCO-India is a private business that supplies PV solar home systems to provide power for lighting and small appliances to low-income households and institutions in Karnataka and Gujarat. SELCO's business model is based on the provision of product and service and providing access to financing through partnerships with banking institutions.[388]

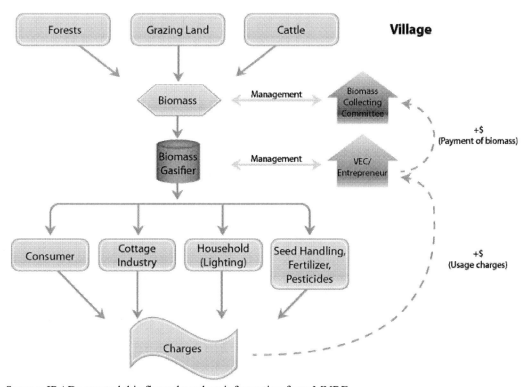

Source: IRADe created this figure based on information from MNRE.

Figure 7-1. Rural energy through biomass gasifier: Business model.

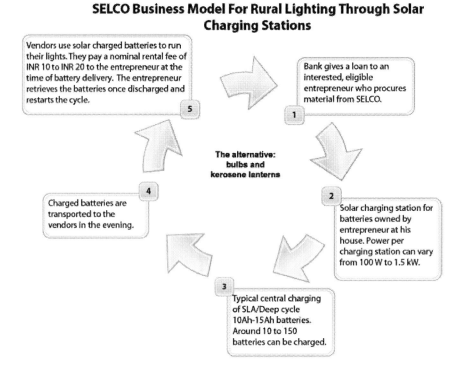

Figure 7-2. SELCO business model[390].

SELCO has supported a business model for PV-powered battery-charging, in which an entrepreneur takes out a loan to purchase a solar lantern charging system and then rents out the solar lanterns to street vendors. Savings realized from no longer needing to purchase kerosene for lanterns are more than the rental costs of the solar lantern (see Figure 7-2).[389]

7.1.3. Micro-hydro Power Systems—A Success Story of Rural Electrification with Renewable Energy in Orissa

Integrated Rural Development of Weaker Sections in India supported development of a 25 kW micro-hydro project in the village of Putsil in the Koraput district of Orissa. This system is designed to supply electricity to surrounding households, the battery-charging station, the grinder/milling machine in the village, and the village community center. The project was commissioned in 1999 after 2 years of intensive interaction with the community.

There is a committee overseeing plant operations, and each family in the community pays a fixed fee for use of the power generated. Members of neighboring communities also pay for use of the mill; half of this revenue stream goes into the community fund, which is to pay for the maintenance of the plant. Along with power generation, the micro-hydro project enhances income-generating activities in the region.[391]

7.2. Captive Power for Industries

According to the Electricity Act 2003, a captive power plant "… means a power plant set up by any person to generate electricity primarily for his own use and includes a power plant set up by any cooperative society or association of persons for generating electricity primarily for use of members of such cooperative society or association." [392]

The opportunities that emerged after the enforcement of the Electricity Act 2003, including de-licensing generation, implementing open access, and setting up a common trading platform, have made the captive power plants an attractive option for industries to meet their in-house requirement and sell surplus power from their captive plants to the grid, which helps the country in meeting power shortage conditions. Apart from the mentioned benefits, others include claiming incentives under CDM and earning RECs, which are going to be launched in India soon.

7.2.1. Captive Power Plants in India

At present, there are about 2,759 industrial units using captive generation power plants with a capacity of 1 MW and above. Most industries supplying their own power are producers of aluminum, cement, chemicals, fertilizers, iron and steel, paper, and sugar, and they use their captive power plants to supplement the electricity purchased from utilities or for emergency use in case of power cuts, restrictions, or failures. As of March 31, 2008, the installed capacity of 1 MW or greater captive power plants in the country was 24,986 MW, as seen in Table 7-1.

These plants generated 90,477 GWh during 2007–2008, which represents a growth of 10.61% from the total 81,800 GWh generated in 2006–2007. Almost all of the generation is from thermal resources, and only very small contributions come from large hydro plants (0.22%) and renewable energy resources (0.42%). Out of a total generation of 90,477 GWh, the energy export to utilities was 7,456 GWh, or 8.2% of the total.

The representation of renewable technologies in the total installed capacity of captive power plants is only 305 MW, or 1.22% of total installed captive power generation capacity. The current capacity is primarily provided by wind turbines located mainly in the southern and western regions of India. The details of captive power installed capacity and generation from wind power plants are given in Table 7-2. [394]

The wind-based captive power plants are owned by different industries, with the textile industry holding the largest share followed by the chemical, iron and steel, light engineering, and cement industries. Not included in these statistics is the bagasse cogeneration of the sugar industry, which is approximately 1,027 MW. Capacity and generation by industry is shown in Table 7-3. [395]

Table 7-1. All India Captive Power Plants of Industry (1 MW and Above)[393]

Source	Installed Capacity (MW)	% share	Gross Generation (GWh)	% share
Hydro	60	0.24	202	0.22
Thermal	24,621	98.54	89,892	99.36
- Steam	- 11,764	- 47.08	- 53,569	- 59.21
- Diesel	- 8,648	- 34.61	- 10,738	- 11.87
- Gas	- 4,209	- 16.85	- 25,585	- 28.28
Renewable Energy	305	1.22	383	0.42
Total	**24,986**	**100**	**90,477**	**100**

Table 7-2. Wind Power Plant Installed Capacity and Generation at Captive Power Facilities

State	Region	Installed Capacity (MW)	Gross Generation (GWh)
Gujarat	Western	75.19	84.82
Maharashtra	Western	16.00	14.16
Chattisgarh		2.50	0
Andhra Pradesh	Southern	2.03	0
Karnataka	Southern	10.14	23.15
Tamil Nadu	Southern	198.89	261.3
Total		**304.75**	**383.43**

Table 7-3. Renewable Energy Installed Capacity and Generation by Industry

Industry	Installed Capacity (MW)	Gross Generation (GWh)
Textile	92.60	159.30
Chemical	46.63	48.61
Iron & Steel	41.0	25.91
Light Engineering	29.59	36.49
Cement	24.47	32.08
Others (Balance)	70.46	81.04
Total (Wind Power)	**304.75**	**383.43**
Sugar	1,027.17	3,317.4

7.2.2. Expansion of Renewable Energy in Captive Power Plants

Captive power generation remains an area of opportunity for several renewable technologies.

7.2.2.1. Wind Power

The industry may continue to set up large capacity wind farms in the states where potential is still available, feed the power into the grid, and utilize equal amounts of power through the state grid at the location of their industry. Alternately, wind power plants can be sited on industry premises. These plants could be developed with suitable storage facilities to meet onsite electric requirements.

7.2.2.2. Solar Power

Industry captive power plants can use solar technologies on rooftops or in parks or other open spaces to meet factories' lighting loads and pumping water requirements.

Additionally, solar thermal power plants can be integrated in existing industries such as paper, dairy, or sugar, which have steam turbines set for cogeneration. Typically these cogeneration units have 5–250 MW capacity and can be coupled with solar thermal power plants. This approach will reduce the capital investment on steam turbines and associated power-house infrastructure, thus reducing the cost of generating solar electricity. The integration of a solar thermal power generation unit with existing coal thermal power plants could be profitable. Savings up to 24% are possible during periods of high insolation for feed water heating to 241° C.[396]

SWHs can serve low-temperature (< 80°) needs for industrial facilities using current technology, but industrial process heat requires up to 250° for advanced technologies. MNRE is supporting R&D of high-temperature process heat powered by solar thermal devices to serve the needs of such industries.[397]

7.2.2.3. Biodiesel

Another area that can increase renewable energy captive power is agricultural plantations of fuel crops, such as Jatropha, used for the production of biodiesel. The industry can use biodiesel to run oil-based diesel generator sets and earn RECs and CDM benefits. However, biodiesel does not yet play a decisive role in captive power generation.

7.2.2.4. Other Sources

Other renewable sources such as ocean waves, ocean thermal energy, and geothermal are site-specific and require a lot of expenditure on surveys, investigation, and R&D. Therefore, these renewable sources are currently not used for captive power plants.

8. TECHNOLOGY TRANSFER

Section Overview

Technology transfer, or the application of internationally developed technologies, is considered to be an important element of a low carbon growth strategy for the Indian economy. Industrialized countries still have a technological lead in a number of renewable energy technologies and it would be desirable—from a global climate change mitigation perspective—to implement them as soon and as widely as possible in India. Transfer of low carbon technologies is considered a key element in combating climate change under the UNFCCC.[398]

Most renewable technologies, perhaps with the exception of solar thermal, geothermal, and offshore wind technologies, already have a strong domestic technology base through Indian companies and international companies in India. India is no longer just a recipient of technologies but increasingly a developer of technologies itself and a strong international competitor in the renewable energy space, as shown in companies such as Suzlon or Moser Baer. Not all renewable technologies are well established in India, but all are being improved upon, and new technologies are continually being commercialized. Much of this research and innovation happens outside India, so there remains a continuous need for technology transfer into India. This can be supported in the public sector through effective capacity building. At the same time, technology transfer from the private sector can be facilitated through an understandable regulatory environment, access to finance, and strong intellectual property protection.

8.1. The Role of the Private and Public Sector in Technology Transfer

A technology transfer typically includes the transfer of the technology design as well as the transfer of the property rights necessary to reproduce the technology in a particular domestic context. A common form of property right included in a technology transfer is a patent license: a legal agreement granting permission to make or use a patented article for a limited period or in limited territory. While the discussion on technology transfer is often held at political levels, it is important to consider that most technologies are not owned by governments but by private corporations. The diffusion of advanced technologies is often driven by multinational companies that want to exploit market opportunities in new regions. From this perspective, technology transfer means the creation of attractive conditions in India for the execution and manufacturing of renewable energy components and applications.

The public sector, which includes the Indian government as well as bilateral and multilateral institutions, plays an important role. Many renewable energy technologies are not yet commercially viable, and in order to thrive, they still depend on incentives given by the public sector.[399] The public sector can also help reduce risks in the market by providing assurances to industries and investors venturing into new or risky technologies or by initiating demonstration projects.

The Indian government, aided by private enterprises and international partners, is also playing an important role in creating relevant educational centers, research hubs, and technology clusters. Furthermore, in a country with such a significant gap in incomes, such as India, it seeks to support the spread of crucial technology to poorer segments of society. This could be called an "internal" technology transfer.

Most importantly, however, the Indian government will decide on the long-term energy strategy and make far-reaching decisions on such issues as the desired mix of energy sources, distribution networks, and electricity pricing.

International platforms such as the international climate change negotiations play an important role as well. They discuss the overall cost associated with climate change mitigation and adaptation and can propose solutions, best practice examples, and financial support or compensation across countries. Moreover, issues such as import tariffs or property rights may be discussed on the international level.

8.2. Technology Transfer through the Private Sector: Opportunities and Barriers

As pointed out in the chapters before, India is an attractive and growing market and is rapidly becoming a key market for renewable energy technologies. The country offers opportunities to investors across the different technologies—wind, small hydro, biomass, solar, and, perhaps soon, emerging technologies such as geothermal and tidal energy—and along the value chain—from component manufacturing to investment, project development, power production, service and maintenance, and training and education.

8.2.1. *Market Opportunities per Technology*

In general, achieving a low price-level is a crucial factor for success of almost all offerings in the Indian market. For the international business community, that will often require localization: setting up a local office or manufacturing unit to become price competitive and understanding the market. Localization could lead to entering into some form of partnership with an Indian company, adapting products and business practices to Indian requirements, and, in the longer term, developing new products in India. India offers great opportunities to international companies to invest in India and bring their technologies to the market.

8.2.1.1. Wind

India has a strong wind market with consistently high growth rates and a long track record, and most wind technologies are locally available. Given a very competitive environment with low and falling costs, the technology may soon reach grid parity.

Being a mature market, the overall growth rate has declined in recent years.[400] There is a shift towards new markets (e.g., new states and offshore) and a need for replacement of older windmills by new and bigger ones.[401]

The growth in the wind market is mainly driven by the still untapped potential and improvements in technology, which lead to increasing capacity factors and turbine sizes combined with decreasing installation cost. Given the current trend towards GB Is, achieving the maximum possible electricity generation becomes crucial. This, in turn, increases the demand for professional project developers. On the manufacturing side, more and more Asian companies are entering the market and thus increasing competition for established players. However, the competitive pressure will be different in various parts of the value chain.

Offshore wind power potential has not been assessed systematically yet. Given India's long coastline and the wind energy potential there, it could play an important role in the future.[402] A transfer of technologies from other countries that already use offshore wind power on a large scale (e.g., Germany) could be possible.

8.2.1.2. Solar

The PV sector is still in the early stages, with an installed grid-connected power capacity of 10.28 MW in March 2010.[403] The long-term potential of solar power in India, given climatic conditions and the rapidly growing electricity requirements, is very attractive. The market is now set to grow significantly and is being driven by JNNSM. As the market begins to pick up speed, there is a particular need for professional EPC contracting and ample supply of silicon wafers and balance-of-system components.[404]

In terms of solar thermal power, there is a clearly discernible market gap in professional EPC services for setting up and running solar thermal power plants. India has very ambitious targets for the development of the technology, yet there are few companies globally who can execute on it.

8.2.1.3. Small Hydro

The development of small hydro (the total potential is estimated at around 15 GW)[405] can contribute to both the national power supply and electrification of remote areas. MNRE's small hydro program currently focuses on reducing capital costs as well as improving reliability and average utilization of facilities.[406]

Small hydro is a well-established technology offering investors projects with a low technology risk and potentially a very attractive return. It currently has on average the lowest generation costs among all renewable energy technologies in India.[407]

8.2.1.4. Biomass

Solid biomass-based power generation (mainly captive cogeneration in the sugar industry) is well established in India. While Indian companies have a lot of experience in combustion of biomass, there still is a significant scope for improving the efficiency of many existing cogeneration plants and technologies.

Biogas has so far been used mainly for small, off-grid applications, but there is a trend towards large-scale applications. New and adaptable technologies are needed as well as logistical, biological, and agricultural expertise. Some Indian and international companies are exploring the technology of upgrading biogas to natural gas equivalent and feeding it into the gas grid.

A great opportunity exists for companies that are able to develop reproducible and sustainable business models and technologies that are applicable in a rural setting. This requires strong distribution channels and a service and maintenance network. Technologies need to be inexpensive, robust, and operational with various locally available feedstock.

8.2.2. Barriers

While being a very promising market for international investors, success in India requires good market knowledge, patience, and adaptability. The regulatory environment is complex and in flux. Products and business models may have to be changed to suit the Indian customer.

8.2.2.1. Product-related Barriers

Many technologies that are well established in developed countries may not be directly transferable to developing nations.[408] Products from outside India may be too expensive due to high production costs in industrialized countries, and product design may not meet the needs of the Indian market or import tariffs. Products often have to be adapted to match market requirements regarding, for example, level of quality, robustness, and climatic conditions. Adaptation may also be needed in business models. To reach competitive price levels, localization of manufacturing, sourcing, or services may be required.[409]

The needs for product adaptation will likely be significantly lower if the technology originates in a country with a market comparable to India's. The Asia Clean Energy Forum, for example, held in Manila in June 2010, sought further cooperation in the renewable energy field from developing nations.[410]

8.2.2.2. Market Information

For foreign companies, a market entry into India is often made difficult by a lack of information about the legal, political, and market environment. The market for renewable energy is very dynamic and, in many sectors, not very transparent. This is partly due to the rapidly changing regulatory environment (such as in the field of solar energy). The federal setup of the Indian state, where individual states formulate policies to complement the Government of India, also plays a role. Especially in technologies such as biomass-based power generation, where local knowledge about land-ownership or agricultural production is essential, information is difficult to obtain.

8.2.2.3. Infrastructure Availability

Another important barrier in many places is the infrastructure.[411] Limited road and grid availability may hamper plans for setting up new plants in remote regions. The hilly areas of the Himalaya, for example, offer huge potential for small hydropower, which is only partially harnessed because of the limited availability of supporting infrastructure.

8.2.2.4. Finance

Financing is a problem in many infrastructure projects where there are high initial costs and a long-term planning horizon. In India, the availability of debt financing is limited due to the perceived risks in the market. These risks relate mostly to PPAs with state power utilities. These have a weak financial position, which raises questions about payment default on long-term PPAs. Also, there is only limited funding—equity and debt—available for new technologies. On the demand side, many Indian consumers may not have access to the capital needed for the higher initial costs of purchasing technologies such as home PV systems.[412]

8.3. The Role of the Public Sector in Improving Technology Transfer

In many sectors, the market for renewable energy is still a politically enabled market with many pre-commercial and supported-commercial technologies, such as solar PV.[413] Therefore, support by the government and bilateral and multilateral institutions and initiatives is needed to enable further development and wider adoption of these technologies. In addition to governmental efforts, bilateral support is provided by industrial nations' development agencies, such as the UK Department for International Development, the German GTZ and KfW, or USAID. Different United Nations organizations (e.g., UNFCCC, World Bank, GEF, and IEA) offer multilateral support and provide international frameworks.

Public sector technology transfer activities will range from needs and resource assessments to selection and adaptation of the appropriate technology to the institutional strengthening required to support technology deployment (see Figure 8-1).

8.3.1. Creating Markets and Providing Incentives

The most important factor limiting the faster spread of clean energy technologies is the relatively higher cost of power generation compared to subsidized alternatives such as fossil fuels.[415] Through a variety of instruments including income tax incentives, preferential tariffs, demand-side measures, and the reduction of support given to fossil fuels,[416] governments can influence the economics in favor of renewable energy.

Another important lever at the government's disposal is import duties and sales taxes. If renewable energy technologies are considered desirable and are supported by public funds, it may be counterproductive to levy high import duties or sales taxes on key components. For example, Finance Minister Pranab Mukherjee announced that the Indian government will exempt renewable machinery, such as solar equipment, parts for rotor blades used in wind turbines, and electric vehicles, from a tax on the production of goods.[417]

In the off-grid market, which is crucial in India, renewable energy solutions do not compete directly with cheap grid power. The government fosters not only international but also inner- Indian technology transfer through creating viable local markets. This requires finding innovative ways to overcome the low purchasing power of consumers relative to the high costs of renewable energy technologies.

The Indian 2010–2011 budget specified some key measures to increase the flow of renewable energy technology into the Indian market. The main points are:[418]

- Establish a National Clean Energy Fund for funding research and innovative projects in clean energy technologies.
- Levy a clean energy tax on coal produced in India (and imported into India) at a rate of INR 50 (USD 1) per ton to build the corpus of a National Clean Energy Fund.
- Increase the weighted deduction on expenditure incurred for in-house R&D from 150%–200% to further encourage R&D efforts. Enhance weighted deductions on payments made to national laboratories, research associations, colleges, universities, and other institutions for scientific research from 125% to 175%.
- Require concessional customs duty provisions of 5% to machinery, instruments, equipment, and appliances for initial set up of PV and solar thermal power generating units and exempt from central excise duty. Exempt ground-source heat pumps used to tap geothermal energy from basic customs duty and special additional duty.
- Complete liberalization of pricing and payment of technology transfer fees and trademark, brand name, and royalty payments.

Figure 8-1. Steps in technology transfer[414].

8.3.2. Support for Financial Markets

The Indian government is looking at improving the financing frameworks of renewable energy projects. It can raise awareness among equity investors and banks regarding the profitability of renewable energy projects. Some investors lack experience and technical understanding. The government assists by providing information and developing standards as well as best practice guides to alleviate the appraisal for developers. For example, MNRE announced guidelines for wind power projects requiring the certification of wind turbines, which reduces the risks for investors.[419]

One of the key levers for improving international investor confidence in the Indian market is through provision of secure, long-term PPAs by the government. Key issues here are the credit status of the contract partners (often the SEBs) and the long-term outlook on feed-in tariffs and electricity prices in India.

Financial support by international donors and innovative financing options are still required in order to scale up investments in renewable energy technologies in India. For example, a renewable energy development project in cooperation by World Bank, GEF, and the Government of India was set up to improve commercial markets and the financing of renewable energy in India. The project promoted and financed various investments of the private sector through IREDA. IREDA carried out market campaigns and business trainings and provided subsidies and financial incentives.[420]

8.3.3. Research and Development and Capacity Building

India wants to become an international leader not only in the use of renewable energy but also in the development and commercialization of new technologies. National policies, such as JNNSM have recognized that in order to achieve more rapid technology development, there is a need for financial assistance for research in India. Funds for research into the manufacturing of polysilicon, for example, are to be provided by the National Energy Fund.[421] The Indian government also plans to set up a Solar Research Council, which would coordinate research activities in India according to the R&D strategy of JNNSM. The strategy includes training, pilot demonstration projects, and National Centres of Excellence as research hubs and "incubators" for innovations and improvement of existing technologies.[422]

There is a clear need for bilateral and multilateral collaboration on R&D of clean energy technologies. Sharing knowledge and experiences can reduce the costs and risks involved in a project. Examples of such cooperation are the Indo-U.S. Science & Technology Forum, which catalyzes the bilateral collaborations in science, technology, and research with the funding vehicle for collaborative research provided under the EU's Framework Program.

Capacity building on the effective use of renewable energy technologies is also required. As a crucial aspect of capacity building, training must be imparted locally. Skills need to be improved to evaluate technological feasibility, cost benefit, applications, and local manufacturing capabilities.

An important example of such work in India is the Bureau of Energy Efficiency's (BEE) Energy Manager Training,[423] which was developed in cooperation with GTZ. BEE publishes a list of certified energy managers and energy auditors on its Web site,[424] and it is clear that this program has successfully created a pool of professionals working towards energy conservation and efficiency.

Support for capacity building is promoted internationally by the Expert Group on Technology Transfer and the UNFCCC,[425] and the United Nations provides direct financial support (e.g., GEF, UNDP, and the World Bank) for trainings, seminars, and workshops.

8.3.4. Legal and Institutional Issues

An important barrier to the introduction of more international technologies in India is the limited enforceability of law. This is particularly relevant with regard to Intellectual Property Rights. Improvements in patent protection could also stimulate domestic innovation. There is a significant increase in resident patent holders for clean technologies in emerging market economies (33% over 1998–2008). In China, approximately 40% of technology patents are locally owned compared to 14% in India. Moreover, climate change technology is not yet as heavily patented as biotechnology, agriculture, and information technology.[426]

8.3.5. Clustering

Through provision of land and tax incentives, the government can encourage the formation of clusters. These could in turn become centers of innovation and business development by concentrating technology specific resources, companies, and market knowledge. Thus, MNRE announced the development of 60 solar cities nationwide, with Nagpur in Maharashtra being the first one. MNRE funds half the costs of the project and requires a 10% reduction of energy consumption through combining improved energy efficiency with greater use of renewable energy sources.[427]

8.3.6. Communication and Information Dissemination

To improve the transfer of renewable energy technologies, an exchange of knowledge and information is required—not only about technical details, but also on the conditions under which a technology is applicable.[428] One example of information dissemination is the special advisory service facilitating technology transfer in clean energy provided by the Asia Pacific Centre for the Transfer of Technology (APCTT).[429] Under their program, "Supporting the Development of an Institutional Cooperation Mechanism to Promote Renewable Energy," APCTT facilitates the collection and dissemination of information on renewable energy technologies, shares best practices on promotion policies and utilization, and promotes R&D collaboration.[430]

Another example is the Indo-German Energy Forum. One of its main objectives is to improve the development of clean energy technologies in the Indian market by promoting transfer of know-how and technology. Topics in the field of renewable energy include support for a solar center at a leading Indian university, piloting solar-energy-based water desalination technology, and developing bankable business models for the use of renewable energy in rural areas. Several other countries are also committed to an exchange of knowledge and technology.

9. FINANCING RENEWABLE ENERGY IN INDIA

Section Overview

Clean energy investment worldwide reached INR 8,100 billion (USD 162 billion) in 2009. This number was down from INR 8,650 billion (USD 173 billion) in 2008 but was still the second highest annual figure ever and nearly four times the total investment in 2004 (INR 2,140 billion/USD 46 billion). The reduction between 2008 and 2009 can be mainly attributed to the recent economic crisis.[431]

Since 1990, India has been part of this boom in investment in renewable energy. According to a recent publication by the United Nations Environment Programme's Sustainable Energy Finance Initiative (SEFI) and Bloomberg New Energy Finance, total financial investment in clean energy in India was at INR 135 billion (USD 2.7 billion) in 2009, ranking it eighth in the world.[432] With an increasingly favorable regulatory and policy environment and a growing number of entrepreneurs and project developers, India has been ranked by Ernst and Young as the fourth most attractive country for renewable energy investment in the world, only behind the United States, China, and Germany.[433]

Asset finance (renewable energy finance projects) and the public market are still the dominant forms of renewable energy financing in India; venture capital and private equity transactions are still limited in India due to the risks associated with renewable energy technologies. The confidence of domestic commercial banks providing loans to renewable projects is still limited; however, this is changing thanks to growing awareness and more and more favorable government policies and targets. Governments and bilateral and multilateral organizations are offering grants at favorable rates to promote renewable energy and project development, which is supporting deployment at all scales. India is also one of the most active participants in the CDM using revenues of carbon credits to finance renewable energy projects.

9.1. Investment by Asset Class

The majority of renewable energy financing in India over the past 5 years has been in asset financing through internal company balance sheets, debt finance, and equity finance. Although the total value of asset financing has fallen from INR 155 billion (USD 3.1 billion) in 2008 to INR 95 billion (USD 1.9 billion) in 2009, it still accounts for 70% of domestic renewable energy investments. Public market activity, money invested in the equity of publicly quoted companies developing renewable energy technology, efficiency, and clean power generation, was around INR 35 billion (USD 700 million) in 2009.[434] The Indian state-owned hydropower company NHPC Limited, raised INR 65 billion (USD 1.3 billion) via an initial public offering on the Indian National Stock Exchange and Bombay Stock Exchange during the year. Indian companies began to explore foreign stock exchanges as an additional source of funds in 2007 and raised INR 37.8 billion (USD 756 million) on the Singapore Stock Exchange and London's AIM in that year, compared to INR 32.3 billion (USD 646 million) on domestic exchanges.[435] However, due to financial turmoil in 2008, those funding avenues dried up with no money raised internationally in 2008 and only INR 3.7 billion (USD 74 million) raised on Indian exchanges.[436] Venture capital and private equity activity in India marked a very small proportion of all investments in 2009 at just 4% of the clean energy total of INR 5 billion (USD 100 million), down from INR 20 billion (USD 400 million) in 2008.[437]

From a technology perspective, the largest portion of new investment went to the wind sector at INR 80 billion (USD 1.6 billion), which represented 59% of the national total and was 11% higher than the investment in wind in 2008. Biomass (including waste-to-energy) received the second largest sector portion of new investment at INR 30 billion (USD 0.6 billion), followed by small hydro and biofuel investments, each at INR 10 billion (USD 0.2 billion).[439] With the release of JNNSM targeting 22 GW of installed solar capacity by 2022 at an estimated cost of approximately INR 27.96 billion (USD 932 million), a substantial increase in investments into solar power is expected.[440] Approval of the 2009 National Biofuels Policy, targeting 20% biodiesel and ethanol blends in fuels by 2017, will also likely trigger an increase in investment.[441]

9.2. Various Types of Financing

Traditionally, the government and NGOs have provided most of the funding for environmental or sustainable development projects in India. The main government funding body has been IREDA, who offers standard loans to renewable energy projects at rates slightly more favorable than general commercial lending rates (see Chapter 2 for more an overview of IREDA's activities).[442,443,444]

As of December 31, 2009, IREDA financed 1,921 projects with IREDA loan commitment amounts totaling over INR 121.8 billion (USD 2.4 billion). Between 1987 and 2009, actual IREDA disbursements are at INR 66 billion (USD 1.41 billion). Such funding has supported the installation of 4.38 GW of power generation capacity. IREDA's loan commitments during 2008 were over INR 14.9 billion (USD 300 million) for 47 projects totaling 403 MW of power generation; 2008 loan disbursements were over INR 7.7 billion (USD 154 million).[445] Other government agencies that actively fund renewable energy projects are the PFC, the Rural Electrification Corporation, and NABARD.[446] Some typical financial products and services are summarized in Table 9-2.

Table 9-1. New Financial Investment in India by Sector and Asset Class, 2009[438]

Sector/Asset Class	Asset Finance		Public Markets		Venture Capital/ Private Equity		Total	
	billion INR	million USD	billion INR	million USD	billion INR	million USD	billion INR	million USD
Solar	2.5	50	1	20	1.5	30	5	100
Biofuels	5	100	5	100	0	0	10	200
Small Hydro	10	200	0	0	0	0	10	200
Biomass and Waste-to-energy	25	500	0	0	5	100	30	600
Wind	50	1,000	30	600	0	0	80	1,600

Table 9-2. Financial Institutions and Services

Name of Institution	Type of Financing
IREDA	Finances up to 70% of project cost with interest rates between 1.75% and 10.50% in the following sectors: 1) Hydro power projects 2) Wind energy: Rebate at 0.25% for a period of 2 years for projects eligible for GBI 3) Biomass cogeneration and industrial cogeneration 4) Solar PV/solar thermal grid-connected power projects 5) Energy conservation/efficiency projects (including DSM) and projects implemented in ESCO mode Finances electrification of remote village projects implemented using ESCO model: 12.75% with term loan up to 80% of project cost New instrument: Loan against securitization of future cash flow of renewable energy projects Implementation agency for finance scheme for off-grid projects under JNNSM
PFC	Provides term loans to all entities (government and IPPs) for power generation (from conventional and renewable energy projects), transmission, and distribution
Rural Electrification Corporation Ltd	Provides loans for all entities (government and independent power producers) for the purpose of rural electrification, irrespective of nature, size, or source of energy Provides short-term loans and debt refinancing to state power utilities in need of financial assistance to cover rural electrification
NABARD	Rural innovation fund for all innovations and related activities in the farm, rural non-farm, and micro-finance sectors that provide technology, skill upgrade, inputs supply, and market support for rural entrepreneurs; assistance is given in the form of loans, grants, or incubation funds Environmental Promotional Assistance: scheme for eco-friendly technologies including biogas, solar, and biofuels Scheme for home lighting through solar energy

The different types of businesses, stages of technology or project development, and degrees of risk associated with investments in renewable energy require different types of investors. At the early stage, when there is no track record and no revenues are generated, the most common form of financing is the entrepreneur's private capital. This normally involves borrowing from friends and family. However, there are also funds available through research and academic institutions, government initiatives, grants and foundations, and within corporate R&D departments.[447] Some identified early stage financiers that are actively supporting renewable energy R&D in India are:

- ICICI Technology Finance Group.
- Techno-Entrepreneurship Promotion Program.
- Indian Institutes of Technology and other universities.
- Corporate R&D departments.
- Centre for Innovation Incubation and Entrepreneurship.
- The Indian Angel Network.

In addition to providing financial support, these financiers may undertake mentoring and capacity building, upgrading and expanding facilities, or bringing in collaborative partners for R&D and business development.

Venture capital is focused on "early stage" or "growth stage" technology companies, while private equity firms tend to look at more mature technologies or projects. They generally expect to exit their investment and make their returns in a 3- to 5-year timeframe.[448] Although Indian venture capital activity is increasing, there are still only a few venture capital investors in India who are actively looking to invest in renewable energy companies. One of the latest deals was from the first quarter of 2010; Azure Power, a solar power plant developer, raised INR 500 million (USD 10 million) in equity funding from IFC, Helion Advisors, and Foundation Capital.[449] Other companies such as Climate Change Capital, Infrastructure Development Finance Company, and FE Clean Energy have created separate funds for clean energy. On the private equity side, with lower cash flows expected in the initial years of renewable energy projects, there is great need for equity funding for construction and expansion. Thus, there appears to be significant scope to attract private equity funds, provided that investment packages are suitably structured and the market offers high, expected returns. Suzlon Energy is an example of a successful private equity investment in renewable energy in India. In 2004, the company attracted private equity investment for funding its expansion plans. Today it is a publicly listed company, having made its initial public offering in 2005.[450]

Corporate financing involves the use of internal company capital to finance a project directly or the use of internal company assets as collateral to obtain a loan from a bank or other lenders. Corporate financiers of renewable energy projects in India are primarily concentrated on large wind and hydropower projects, where captive power generation and the application of accelerated depreciation benefits play a significant role.[451] Project financing is a method of financing a project through debt and equity that are repaid by the revenues and assets of a specific project rather than through the balance sheet of a company. The primary debt providers in renewable energy project financing are commercial banks. For a long time, Indian domestic banks were shy to lend to the sector. There were perceived technological risks, risks related to PPAs, and risks associated with escalating project costs. However, as the market matured and more favorable government policies passed, banks have become increasingly more comfortable funding renewable energy projects at attractive interest rates.

Prominent domestic banks that currently fund renewable projects are: Industrial Development Bank of India (IDBI), Export-Import Bank of India, ICICI Bank, the Industrial Finance Corporation of India (IFCI), State Bank of India, Yes Bank, and PNB. ICICI Bank, India's largest private sector lender, has been funding renewable energy projects for years. Currently, ICICI provides a 50% waiver on loan processing for green vehicles and green homes.[452] Yes Bank, another example, has set aside nearly 20% of its infrastructure to lend to fund clean technology projects; it recently approved a loan of INR 1 billion (USD 20 million) for wind farm equipment supplied by Enercon, which operates under the independent power producer model.[453]

In the so-called "bottom of the pyramid" market, a number of microfinance institutions (MFIs) provide credit to individual borrowers or to local NGOs to give them access to responsive and timely financial services at market rates, to repay their loans, and to use the proceeds to increase their income and assets. MFIs facilitate the purchase of renewable energy systems like solar cookers, solar lanterns, or small biogas plants in off-grid areas of the country. The Self Employed Women's Association is perhaps the most well-known example of an MFI in India. Also, a number of government organizations and private banks provide financial assistance to small- and medium-sized enterprises (SMEs). The various financial services and products provided in the field of microfinance and SME finance are summarized in Tables 9-3 and 9-4.

9.3. International Support

The Indian renewable energy sector has received significant support from the international community in the form of grants and low-interest loans. Several international financial institutions and bilateral financial institutions are engaged in renewable energy financing in India, such as World Bank/IFC, KfW, Nordic Investment Bank, United Nations Environment Programme (UNEP), ADB, USAID, and Danish International Development Agency (DANIDA). Their activities are summarized in Table 9-5, some of which are discussed more in detail below.

Table 9-3. Commercial Banks and their Services

Indian Commercial Banks	Type of Financing
Bank of Baroda	Infrastructure financing
Bank of Maharashtra	Infrastructure financing Energy equipment financing (solar energy, bioenergy, and clean energy programs)
Canara Bank	SME financing Infrastructure financing End-user financing
Corporation Bank	Infrastructure financing
Export - Import Bank of India	Technology financing Infrastructure financing
ICICI Bank	Infrastructure financing (green homes) Transport financing (green cars)
IDBI Bank	Carbon credit business SME financing Infrastructure financing
IFCI	Venture Capital (acquired stake in Luminous Teleinfra Limited)
Indian Overseas Bank	Infrastructure financing
ING Vysya Bank Ltd.	Infrastructure financing
Laxmi Vilas Bank	Infrastructure financing
Punjab National Bank	Infrastructure financing
State Bank of Bikaner and Jaipur	Infrastructure financing
State Bank of India	Infrastructure financing
Axis Bank	Infrastructure financing Private equity (acquired a minority stake in a south India clean energy firm)

Table 9-4. Microfinance Institutions and Initiatives

Institution Name	Type of Microfinance
Aryavart Gramin Bank	Approved loans for the installation of 8,000 solar-home systems in Uttar Pradesh
Grameen Surya Bijlee Foundation	Helps fund solar lamps and home and street lighting systems for villages in India, Nepal, and Bangladesh
Green Microfinance and MicroEnergy International launch Energizing India, with the help of the Evangelical Social Action Forum	Provides micro-energy products for its clients (micro-businesses and families) in four provinces in south India through micro-loans
HSBC and Micro Energy Credits	Provides clean energy alternatives to clients of Spandana (enables them to access the global carbon credit market)
Renewable Energy and Energy Efficiency Partnership (REEEP), TERI, and Clinton Climate Initiative[454]	Replaces kerosene and paraffin lanterns with solar devices
REEEP[455]	Developing 10 renewable energy projects with microfinancing
Self Employed Women's Association	Provides affordable renewable energy sources to poor people, e.g., solar-powered batteries
SKS Microfinance	Offers solar lamps (to women)

9.3.1. World Bank/International Finance Corporation

The World Bank's annual lending to India's power sector is approximately INR 50 billion (USD 1 billion). The World Bank is supporting the efficient transmission and distribution of power to consumers; in September 2009, they extended a loan of INR 50 billion (USD 1 billion) to Powergrid to strengthen and expand five transmission systems in the northern, western, and southern regions of the country, and they are also supporting transmission and distribution improvements in Haryana and Maharashtra. Through the Bank's private sector arm, the IFC, a loan of INR 312 million (USD 6.25 million) was approved for renewable energy company Auro Mira Energy Company Private Limited to build an 18 MW biomass- based power plant.[456] In 2010, IFC also provided a INR 3.8 billion (USD 75 million) loan to Infrastructure Development Finance Corporation (IDFC) for investments in renewable energy, cleaner production, and energy efficiency projects.[457]

9.3.2. KfW

Since the 1950s, more than EUR 8 billion (INR 480 billion/USD 9.6 billion) have been channeled by KfW on behalf of the German government, mainly for the fields of energy, finance sector development, health, and protection of the environment and natural resources. Currently, 12 energy sector projects, with KfW loan commitments of more than EUR 1.2 billion (INR 70 billion/USD 1.4 billion), are being implemented in India. In 2009, KfW signed an agreement with IREDA to provide financial assistance of EUR 19.97 million (INR 1.20 billion/USD 24.00 million) for the promotion of biomass power generation in India. The loan aims to accelerate the adoption of environmentally sustainable biomass power and cogeneration technologies in India and promotes innovative biomass projects of combustion, gasification, methanation, and cogeneration technologies for electricity generation.[458]

9.3.3. United Nations Environment Programme

The 4-year, INR 380 million (USD 7.6 million) India Solar Loan Program was launched in 2003 through a partnership between UNEP, the UNEP Risø Centre on Energy, Climate and Sustainable Development, and two of India's largest banking groups, with support from the UN Foundation and Shell Foundation. It aimed to establish a consumer credit market for financing solar home systems in southern India. The innovative financing arrangement involved an interest rate reduction, market development support, and a process to qualify solar suppliers. The interest rate reduction was phased out during the program and after project completion. Further financing of solar home systems is on purely commercial terms. While the solar home sector was a small, cash-only business in 2003, today the market is growing with more than 50% of sales financed by banks. There are now more than 20 banks with networks of more than 2,000 branches offering solar financing.[459]

9.3.4. Asian Development Bank

ADB's energy sector program includes support for enhancing the impact of the Government of India's initiatives in the renewable energy sector. There are many ongoing ADB lending programs targeting various renewable energy activities in different states in India. For instance, ADB is providing almost INR 7.5 billion (USD 150 million) in loans to support five wind power projects totaling over 260 MW in Maharashtra, Gujarat, and Karnataka. In line with its Asian Solar Energy Initiative, ADB is presently supporting MNRE in implementation of JNNSM, and planned activities include support for the development of large-scale (1–3 GW) solar power parks in the states of Gujarat and Rajasthan. ADB's private sector operations is also preparing an INR 7.5 billion (USD 150 million) guarantee facility to mobilize commercial financing for an aggregate 100 MW of small grid-connected solar projects and is also evaluating direct lending for larger solar PV and solar thermal projects in India.[460]

9.4. Clean Development Mechanism[461]

The first phase of Kyoto Protocol aims at bringing the GHG levels in the atmosphere to 5.4% below the 1990 level by the year 2012. UNFCCC has identified three mechanisms under the Kyoto Protocol to enable the reduction of GHG emissions in a cost-effective approach. CDM is one of the flexible mechanisms by which this can be achieved through GHG emissions trading between the Annex I countries (mainly industrialized countries) and Non-Annex I countries under the climate change convention.[462] Financing is one of the major hurdles faced by emission reduction projects and the primary goal of the CDM is making financing available to such projects to encourage sustainable development. With the provision of the CDM principle, India could sell achieved emission reductions credits to developed countries.

Table 9-5. Bilateral and Multilateral Institutions and Services

Name of Institution	Type of Financing
ADB	• Private sector financing—equity, loans, and guarantees • Technical Assistance Special Fund
World Bank and IFC	• Infrastructure financing (IFC)—first commercial utility solar project in India to Azure Power Private • Microfinancing—INR 453 million (USD 9.05 million) grant to Small Industries Development Bank of India • Private sector funding • Renewable energy financing and commercial market improvements (with Global Environment Facility); the project allowed IREDA to promote and finance private-sector renewable investments • World Bank and USAID loan disbursement by ICICI Bank
Global Environment Facility (GEF)	• Private equity • Private sector funding • Renewable energy financing and commercial market improvements (with GEF); the project allowed IREDA to promote and finance private-sector renewable investments • Special fund to promote PV market transformation
KfW	• Lines of credit to Indian finance institutions—IREDA, PFC, REC • Technology financing • Microfinancing • SME financing
DANIDA	• Infrastructure financing—wind power • Soft loans • Technology financing
USAID	• Venture capital fund—initiated by Indian financial institutions to support clean technology activities in the private marketplace • Infrastructure financing • World Bank and USAID loan disbursement by ICICI Bank
Nordic Investment Bank	• Line of credit of INR 2.5 billion (USD 50 million) for financing of renewable energy projects

India began participation in the CDM in 2003, and as an early entrant has been an active and significant participant using revenues of carbon credits to finance renewable energy projects. India has the second largest market share of CDM projects after China; the CDM executive board had registered 506 projects in India as of June 2010, which amounted to more than 20% of all the registered projects.[463] In the initial stage of CDM development in India, biomass utilization projects, waste gas/heat utilization projects, and other renewable energy technology projects (largely wind and hydropower) were mainly being implemented, though India also has registered CDM projects for other sectors including energy efficiency, afforestation and reforestation, and transportation. Figure 9-1 shows the contributions of various project categories to India's CDM portfolio.

Indian Renewable Energy Status Report: Background Report for DIREC 2010

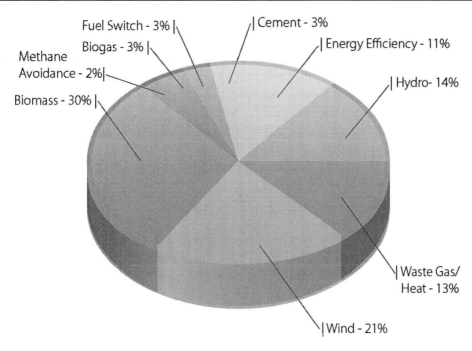

Figure 9-1. Types of registered CDM projects in India[464].

There are also a large number of unilateral CDM projects, approximately 422 projects as of June 2010,[465] which are being developed by Indian stakeholders without the technological or financial engagement of Annex I industrialized countries. Indian project developers implement the project by the bearing transaction costs of CDM and taking on the project risks.

In the past few years, a large number of renewable energy projects have benefited from the CDM process. With the development of programmatic and sectoral CDM, it is expected that more small-scale renewable energy projects will be bundled together to tap the revenue from international emission trading.

10. ENABLING ENVIRONMENT

Section Overview

Future success in achieving ambitious renewable energy goals will require supportive policies, innovative financing availability, and technology growth. The International Renewable Energy Conference (IREC) series brings together stakeholders to discuss the conditions that have supported successful renewable energy deployment, the challenges still being faced, and creative solutions. An increasing number of countries are setting targets for renewable generation at the national and sub-national levels. In addition and as an alternative, countries are setting preferential tariffs to deploy technologies that cannot yet compete on costs with traditional fossil technologies. Governments are increasing their fiscal support for renewable energy in a variety of ways including funds for demonstration projects and loan guarantees. Public sector international financial institutions are playing a larger role in supporting clean energy projects relative to private financial institutions, though the latter are still active participants and may increase their role again as global financial markets recover. At the same time that renewable energy is receiving this policy and fiscal support, the technologies themselves are continually improving and the costs coming down. So as India and other nations gather for DIREC 2010 and look ahead to the next gathering in the series, there are many opportunities to share experiences that have contributed to renewable energy growth to-date and prospects for more rapid growth in the coming years.

10.1. History of the International Renewable Energy Conference Process

Since the initiation of the first IREC in Bonn, Germany, in 2004, this conference series has achieved high-level, international participation of stakeholders from policy, business, science, and civil society. The Bonn "Renewables 2004" and the subsequent Beijing International Renewable Energy Conference (BIREC 2005) and Washington International Renewable Energy Conference (WIREC 2008) provided important inputs for the international discourse on renewable energy.

Renewables 2004 charted the way towards an expansion of renewable energy technologies worldwide, responding to the call of the Johannesburg summit for the global development of renewable energy in 2002. It also kept up the momentum generated by a group of like- minded countries for promotion of renewable energy. Renewables 2004 addressed the issue of how the proportion of renewable energy can be substantially increased and how its advantages and potential can be better used.

The conference concentrated in particular on the following themes:

- Formation of enabling political framework conditions, allowing the market development of renewable energy.
- Increase in private and public financing in order to secure reliable demand for renewable energy.
- Human and institutional capacity building and coordination and intensification of R&D.

BIREC in November 2005 called on the world to consider renewable alternatives in the era of high oil prices. The political commitment to encourage the use of fossil fuel alternatives was strengthened by the messages from the Chinese President and the General Secretary of the United Nations to the 1,200 representatives from more than 80 countries and regions.

WIREC was held March 4–6, 2008, in Washington, D.C. WIREC was a collective effort, involving public and private partners, both domestic and international. Building on the two preceding conferences, WIREC again brought together the world's leaders in the field of renewable energy to reconfirm the commitment to a cleaner energy future. In partnership with REN21, WIREC gathered 145 specific, policy-oriented initiatives that will significantly increase the presence of renewable energy worldwide.

DIREC 2010, with special focus on "Upscaling and Mainstreaming Renewables for Energy Security, Climate Change, and Economic Development," is the fourth in the series of the global ministerial-level conferences on renewable energy. The Government of India hosts this international platform for government, private sector, and civil society leaders to jointly address the goal of advancing renewable energy. It will build upon the success and outcomes of the previous events in Bonn, Beijing, and Washington, D.C.

10.2. International Policy Landscape

Currently, 83 countries promote the use of renewable energy resources through national and regional policies, and this number continues to increase. These policies include feed-in tariffs, renewable obligations or quotas, capital subsidies or grants, investment tax credits, sales tax or value added tax exemptions, green certificate trading, direct energy production payments or tax credits, net metering, direct public investment or financing, and public competitive bidding. Apart from the national level, local governments and even cities play a role. [466]

The effectiveness of the policies depends on the governments' abilities to commit towards renewable energy in the long term, to install legally binding targets, and to improve while continually building on past experiences. To enhance the impact, most countries use a combination of different policies at the national, state, and regional levels.

As Asian countries improve their policies on and commitment for renewable energy, the leadership in this market is shifting from West to East, especially to China who added the largest renewable energy capacity of 37 GW in 2009 and to India. [467]

10.2.1. Policy Targets

National policies can set targets for the share of renewable energy in total electricity, primary or final energy mix. The first period for targets ends in 2012, but new policies aim at a longterm commitment until 2020 and beyond. [468]

An example for such a policy is the agreement of the 27 countries of the EU to follow an EUwide target. The EU stated that by 2020, 20% of their final energy mix should be produced by renewable energy resources. [469] In some countries, such as in the United States and Canada, targets are set on a state or regional level. [470]

In the case of countries such as India who have double-digit growth rates in energy demand, [471] setting percent targets is ambitious. Nonetheless, both countries have set themselves targets. India's Eleventh Plan sets a renewable energy capacity target of 25 GW by 2012, and JNNSM aims for installed on-grid solar alone to reach 20 GW by 2022. [472]

The policy targets are communicated and discussed on an international level during global conferences such as DIREC. The International Renewable Energy Agency (IRENA) will support the efforts of different countries to achieve their stated targets. [473]

10.2.2. Renewable Energy Promotion Policies

The most common promotional policies are preferential feed-in tariffs. They are implemented in 50 countries and have been effective at speeding up the deployment of renewable energy technologies, particularly in the wind sector. [474]

In recent years, many countries have modified their tariffs based on prior experience and changes in the market. China, for example, based its new tariffs for wind power on competitive bidding and project development experience. Furthermore, it developed a system of different tariffs for different regions of the country based on the availability of renewable energy resources.[475] In some countries, such as Spain, the reduction of technology costs and growing global competition led to the announcement of lower tariffs.[476]

National policies for quotas on renewable energy or renewable portfolio standards exist only in 10 countries (Australia, Chile, China, Italy, Japan, the Philippines, Poland, Romania, Sweden, and the United Kingdom), but these are more common on the state/provincial level, especially in the United States and India. Most of these policies target a share of renewable energy power in the range of 5%–20%.[477]

Direct capital investment subsidies or grants are offered either for a broad range of renewable energy technologies, as in Indonesia and the Philippines, or are technology specific, as in India's income tax exemption for wind projects. Recently, the import duties were reduced in many countries to remove this barrier to the growth of renewable energy. An example is South Korea, where duties have been reduced by 50%.[478]

The solar PV market, especially PV rooftop programs, is often supported by capital subsidies, which usually cover 30%–50% of the installation costs. In South Korea, as much as 70% of the cost is subsidized and a goal of 300 MW from rooftop PV by 2011 was announced. In China, half of the capital costs of building-based solar PV larger than 50 kW and on-grid projects of more than 300 kW are subsidized.[479]

Tradable RECs are well established in the EU and in some U.S. states. The European Energy Certificate System consists of 18 countries and provides a platform for the issue and transfer of voluntary RECs as well as guarantee-of-origin certificates.[480]

Public investment, low-interest loans or financing is considered to be an important aspect for growth in the renewable energy sector. Therefore, many countries have established renewable energy funds. One of these is the Canadian CAD 1 billion clean energy fund (INR 45 billion/USD 987 million) that supports the demonstration of projects as well as R&D.[481] New funds have come up in China, the Philippines, Bangladesh, and Jordan. In India, the establishment of a National Clean Energy Fund is part of the Indian Budget 2010–2011.[482]

Public competitive bidding policies involve bidding for fixed quantities of renewable energy capacity. One of the oldest examples of this policy, China's bidding policy for wind projects, is being replaced by feed-in tariffs. American countries, such as Brazil, Uruguay, and Argentina, launched competitive bidding programs in 2008–2009.[483]

10.3. New Financing Methods

The total investment in renewable energy capacity in 2009 was INR 7,500 billion (USD 150 billion). Venture capital and private equity investment went down mainly as a result of the global financial crisis. At the same time, governments increased the investment in research, development, and deployment. Germany and China were the investment leaders in 2009, each spending around INR 1,250-1,500 billion (USD 25–30 billion). Wind power received 62% of the total investment in utility-scale power plants. Investments in PV declined, mainly due to falling module prices.[484]

An interesting model to meet the demand for risk-mitigation is the insurance for CDM projects. The insurers guarantee that CERs will be delivered on agreed terms and conditions to the client or they will provide financial compensation. This reduces the risk for project developers, investors, and buyers of CERs. Institutions offering this service are the Austrian Garant Insurance, the French Global Sustainable Development Project, and Swiss Re Greenhouse Gas Risk Solutions. The first insurance was implemented in South America and covers a carbon delivery guarantee, political risk insurance, and business interruption.[485]

The last years have seen significant "green stimulus packages" that include grants, loans, loan guarantees, tax credits, tax deduction, and other supporting schemes. With INR 3,345 billion (USD 68.7 billion), China leads the investments, followed by the United States with INR 3,330 billion (USD 66.6) billion and EU with INR 565 billion (USD 11.3 billion). The green stimulus funding has so far suffered from delays in the administrative processes. To bring the funds to bear more quickly, they can be injected by development banks. The Japan Bank of International Cooperation, for example, has set up an INR 250 billion (USD 5 billion) fund for environmental projects in Asia. The French development bank, Agence Française de Dévellopement, plans to invest INR 520 billion (USD 10.4 billion) in infrastructure projects.[486]

While public sector banks and development banks expanded their roles, the commercial banks reduced their exposure. The most important players include the European Investment Bank, the Brazilian National Bank of Economic and Social Development, the World Bank Group, the German KfW, the Inter-American Development Bank, and ADB.[487]

10.4. Renewable Technology Costs

Costs for many renewable energy technologies have been falling significantly over the last years, making them more competitive. A good example is solar PV; in the last 12 years, the module costs, which account for about 50% of total costs, declined due to economies of scale in production. At the same time, non-module costs such as inverters, other hardware, labor, permitting, and shipping also came down.[488]

A main component in the production of PV crystalline cells is polysilicon. The supply of material was constrained and prices doubled from 2004 until 2008. This is not considered a problem, as improvements in utilization and new production capacities lead to the assumption of falling future costs. Regarding thin-film PV, a potential constraint and cost driver could be the supply of some rare metals used for this technology.[489]

10.5. New Technology Development

In the coming years, we cannot only expect the drops in renewable technologies discussed above, but we can also expect substantial progress with some of the less mature renewable technologies. Some of these technologies are ready to see significant gains, and some may see their utilization rates accelerated by ambitious initiatives of governments and the private sector.

10.5.1. Geothermal Power

Geothermal power plants provide direct heat and electricity. Twenty-four countries currently have geothermal power plants with a total capacity of 10.7 GW, but 88% of it is generated in just seven countries: the United States, the Philippines, Indonesia, Mexico, Italy, New Zealand, and Iceland. The most significant capacity increases since 2004 were seen in Iceland and Turkey. Both countries doubled their capacity. Iceland has the largest share of geothermal power contributing to electricity supply (25%), followed by the Philippines (18%).[490] The development of advanced technologies is expected in the near future, as well as a rise in the number of installations. Currently, there are many new projects under construction.[491]

10.5.2. Ocean Energy

To date, ocean energy is mostly used in pilots and demonstration projects. While tidal and wave technologies are already implemented commercially in a few projects, ocean thermal energy conversion is still in exploration status. The governments of some European states, Australia, South Korea, Canada, and the United States support the growth of the market by grants. Ireland has the largest and most established market with preferential feed-in tariffs for ocean energy.[492]

10.5.3. Large Grid Networks

The concept of large grid networks is being promoted by the DESERTEC initiative, which promotes deployment of CSP in desert regions and transport the electricity generated using high-voltage direct current transmission networks. They are supporting CSP development in North Africa with an aim to provide most of the power needs for the Middle East and North Africa and up to 15% of the power needs of Europe by 2050. The initiative supports use of a large network of medium transmission lines to move power from North Africa to Europe instead of a few large transmission lines (analogous to the transmission networks for oil and natural gas) to increase security of supply.[493]

10.5.4. Solar Cooling

Use of solar thermal systems to supply energy to absorption chillers is being explored as part of several ambitious renewable energy initiatives. The Masdar City project in the United Arab Emirates and the Hawaii Renewable Energy Deployment venture in the United States have both announced plans to utilize the technology to cool buildings.[494,495] Solar cooling technology presents opportunities for many warm climates, including India, where a solar thermal cooling system has already been deployed at the Muni Seva Ashram in Gujarat.[496]

10.6. Looking Ahead

As policymakers and renewable energy professionals convene at DIREC, they will be looking at successes since the last IREC in Washington, D.C., and forward to what can be achieved before the next meeting in 2012. Country- and state-level adoption of targets for renewable capacity and generation has increased rapidly in recent years, renewable technologies continue to improve, and deployment of these technologies is continuing in a way that their contribution to the global energy system continues to increase in significance. Policy action in countries like India has resulted in the contribution of increasing renewable energy market growth in developing countries. At the next IREC, India will be poised to present on lessons learned from the first phase of JNNSM and plans for Phase 2, and there may be changes to the Indian manufacturing sector to reflect the industry growth objectives of the mission. If current trends continue, the next IREC will take place amidst a rapid expansion of installed renewable capacity with delivered energy costs continuing their trend toward parity with fossil technologies.

REPORT DOCUMENTATION PAGE		Form Approved OMB No. 0704-0188

The public reporting burden for this collection of information is estimated to average 1 hour per response, including the time for reviewing instructions, searching existing data sources, gathering and maintaining the data needed, and completing and reviewing the collection of information. Send comments regarding this burden estimate or any other aspect of this collection of information, including suggestions for reducing the burden, to Department of Defense, Executive Services and Communications Directorate (0704-0188). Respondents should be aware that notwithstanding any other provision of law, no person shall be subject to any penalty for failing to comply with a collection of information if it does not display a currently valid OMB control number.
PLEASE DO NOT RETURN YOUR FORM TO THE ABOVE ORGANIZATION.

1. REPORT DATE *(DD-MM-YYYY)* October 2010	2. REPORT TYPE Technical Report	3. DATES COVERED *(From - To)*
4. TITLE AND SUBTITLE Indian Renewable Energy Status Report: Background Report for DIREC 2010		**5a. CONTRACT NUMBER** DE-AC36-08-GO28308
		5b. GRANT NUMBER
		5c. PROGRAM ELEMENT NUMBER
6. AUTHOR(S) D. S. Arora (IRADe); Sarah Busche (NREL); Shannon Cowlin (NREL); Tobias Engelmeier (Bridge to India Pvt. Ltd.); Hanna Jaritz (IRADe); Anelia Milbrandt (NREL); Shannon Wang (REN21 Secretariat)		**5d. PROJECT NUMBER** NREL/TP-6A2-48948
		5e. TASK NUMBER IGIN.9700
		5f. WORK UNIT NUMBER
7. PERFORMING ORGANIZATION NAME(S) AND ADDRESS(ES) National Renewable Energy Laboratory 1617 Cole Blvd. Golden, CO 80401-3393		**8. PERFORMING ORGANIZATION REPORT NUMBER** NREL/TP-6A2-48948
9. SPONSORING/MONITORING AGENCY NAME(S) AND ADDRESS(ES)		**10. SPONSOR/MONITOR'S ACRONYM(S)** NREL
		11. SPONSORING/MONITORING AGENCY REPORT NUMBER
12. DISTRIBUTION AVAILABILITY STATEMENT National Technical Information Service U.S. Department of Commerce 5285 Port Royal Road Springfield, VA 22161		
13. SUPPLEMENTARY NOTES		
14. ABSTRACT *(Maximum 200 Words)* India has great potential to accelerate use of endowed renewable resources in powering its growing economy with a secure and affordable energy supply. The Government of India recognizes that development of local, renewable resources will be critical to ensure that India is able to meet both economic and environmental objectives and has supported the development of renewable energy through several policy actions. This paper describes the status of renewable energy in India as of DIREC 2010. It begins by describing the institutional framework guiding energy development in India, the main policy drivers impacting energy, and the major policy actions India has taken that impact renewable energy deployment. The paper presents estimates of potential for wind, solar, small hydro, and bioenergy and the deployment of each of these technologies to date in India. The potential for India to meet both large-scale generation needs and provide access to remote, unelectrified populations are covered. Finally, the enabling environment required to facilitate rapid scale of renewables is discussed, including issues of technology transfer and the status of financing in India.		
15. SUBJECT TERMS India; renewable energy; wind; solar photovoltaics; PV; bioenergy; biogas; biofuels; biomass; hydropower; solar water heating; concentrating solar power; climate change; finance		

16. SECURITY CLASSIFICATION OF:			17. LIMITATION OF ABSTRACT	18. NUMBER OF PAGES	19a. NAME OF RESPONSIBLE PERSON
a. REPORT Unclassified	b. ABSTRACT Unclassified	c. THIS PAGE Unclassified	UL		19b. TELEPHONE NUMBER *(Include area code)*

End Notes

[1] European Commission. "World Energy, Technology and Climate Policy Outlook." http://ec.europa.eu/research/energy. Accessed October 2010.

[2] IEA. "World Energy Outlook 2008." http://www.worldenergyoutlook.org/docs/weo2008/WEO2008.pdf. Accessed October 2010.

[3] Public Diplomacy Division, Ministry of External Affairs. "The Road to Copenhagen: India's Position on Climate Change Issues." Government of India. http://pmindia.nic.in/Climate%20Change 16.03.09.pdf. Accessed May 2010.

[4] Ministry of Environment and Forests. *Annual Report 2009-2010.* p. 271. http://moef.gov.in/report/report.html. Accessed August 2010.

[5] See the recent study: Center for Science and Environment (CSE). "Challenge of the New Balance." April 2010. IRADe is currently performing an even more comprehensive study on India's low carbon growth options for the National Planning Commission of India.

[6] Ministry of Environment and Forests. *Annual Report 2009-2010.* p. 268. http://moef.gov.in/report/report.html. Accessed August 2010.

[7] Press Information Bureau, Government of India. "Ministry of Environment and Forests." http://pib.nic.in/release/rel_print_page1.asp?relid=44098. Accessed September 2010. For further information, see: http://pmindia.nic.in/Pg01-52.pdf. Accessed September 2010.

[8] UNFCCC. "Clean Development Mechanism—About CDM." http://cdm.unfccc.int/about/index.html. Accessed May 2010.

[9] Ministry of Environment and Forests. *Annual Report 2009-2010.* pp. 269-270. http://moef.gov.in/report/report.html. Accessed August 2010.

[10] CDM. "Project Search." http://cdm.unfccc.int/Projects/projsearch.html. Accessed July 2010.

[11] Indiacore. "Overview of Indian Power Sector." http://www.indiacore.com/power.html. Accessed August 2010.

[12] McKinsey. "Powering India: The Road to 2017." 2009 (unpublished).

[13] BP. "Statistical Review of the World Economy, 2009." http://www.bp.com/liveassets/bp_internet/globalbp/globalbp_uk_english/reports_and_publications/statistical_en ergy review 2008/STAGING/local assets/2009 downloads/statistical review of world energy full report 2 009.pdf. Accessed September 2010. Calculation based on the figures of 2000–2008.

[14] Government of India Planning Commission. "Integrated Energy Policy—Report of the Expert Committee 2006." August 2006; pp. 18-19. http://planningcommission.nic.in/reports/genrep/rep intengy.pdf. Accessed September 2010.The study indicates that the international average value is 0,83%, whereby countries with a higher GDP (over USD 8,000 per person per year) tend to have a lower average energy elasticity of 0.76% (see p. 19).

[15] Government of India Planning Commission. "Integrated Energy Policy—Report of the Expert Committee." August 2006; p. 13 (overview). Online http://planningcommission.nic.in/reports/genrep/rep intengy.pdf. Accessed September 2010.

[16] Government of India Planning Commission. "Integrated Energy Policy—Report of the Expert Committee." August 2006; p. 20. Online http://planningcommission.nic.in/reports/genrep/rep_intengy.pdf. Accessed September 2010.

[17] McKinsey. "Powering India: The Road to 2017." 2009 (unpublished).

[18] McKinsey. "Powering India: The Road to 2017." 2009 (unpublished).

[19] Commercial energy resources contain all the commercially traded fuels. Although they play a major role in many countries, wood, peat, and animal wastes are not included because no reliable statistics are available.

[20] BP. "Statistical Review of World Energy, 2009." http://www.bp.com/liveassets/bp_internet/globalbp/globalbp_uk_english/reports_and_publications/statistical_en ergy review 2008/STAGING/local assets/2009 downloads/statisticalreviewofworldenergyfullreport2009.pdf. Accessed September 2010.

[21] BP. "Statistical Review of World Energy, 2009." http://www.bp.com/liveassets/bp internet/globalbp/globalbp uk english/reportsandpublications/statisticalenergyreview2008/STAGING/localassets/2009downloads/statistical review of world energy full report 2 009.pdf. Accessed September 2010.

[22] Census of India. "2001 Data Highlights." http://censusindia.gov.in/Data_Products/Data_Highlights/ Data_Highlights_link/data_highlights_hh1_2_3.pdf. Accessed October 2010.

[23] Census of India. "Fuel Used for Cooking." http://censusindia.gov.in/Census_Data_2001/Census_data_finder/HH_Series/Fuel_used_for_cooking. Accessed October 2010.

[24] Government of India Planning Commission. "Integrated Energy Policy—Report of the Expert Committee." August 2006; p. 29. Online http://planningcommission.nic.in/reports/genrep/rep intengy.pdf. Accessed September 2010

[25] CEA. http://www.cea.nic.in/. Accessed September 2010.

[26] Power Line. "Impetus to T&D." June 2009.

[27] World Economic Forum. "The Global Competitiveness Report 2010-2011." Geneva, Switzerland, 2010; p. 394; http://www3.weforum.org/docs/WEF_GlobalCompetitivenessReport_2010-11.pdf. Accessed September 2010.

[28] CEA. "Power Sector Reports." http://www.cea.nic.in/power sec reports/reports.htm. Accessed August 2010.

[29] Wärtsilä. "The Real Cost of Power." 2009. http://www.war. Accessed September 2010.

[30] O'Neill, J.; Poddar, T. "Global Economics Paper No: 169." Goldman Sachs. http://www2.goldmansachs.com/ideas/brics/ten-things-doc.pdf. Accessed September 2010.

[31] Sumir Lal. "Can Good Economics Ever Be Good Politics: Case Study of India's Power Sector." World Bank Working Paper No. 83. 2006; p. 16.

[32] Sumir Lal. "Can Good Economics Ever Be Good Politics: Case Study of India's Power Sector." World Bank Working Paper No. 83. 2006; pp. 3-5.

[33] CEA. "Monthly Review of Power Sector (Executive Summary)" August 2010. http://www.cea.nic.in/power_sec_reports/executive_summary/2010_08/1-2.pdf. Accessed October 2010.

[34] CEA. "Monthly Review of Power Sector (Executive Summary)" August 2010. http://www.cea.nic.in/power sec reports/executive summary/2010 08/1-2.pdf. Accessed October 2010.

[35] CEA. "All India Electricity Statistics—General Review." New Delhi: CEA. 2009.

[36] CEA. "Growth of Electricity Sector in India from 1947-2009." New Delhi: CEA, 2009.

[37] India Core. "Overview of Power Sector in India, 2008." New Delhi: India Core Publishing, 2008.

[38] McKinsey. "Powering India: The Road to 2017." 2009 (unpublished).

[39] CEA. "Monthly Review of Power Sector (Executive Summary)." March 2010. http://www.cea.nic.in/power_sec_reports/executive_summary/2010_03/1-2.pdf. Accessed October 2010.

[40] CEA. "Monthly Review of Power Sector (Executive Summary)" March 2010. http://www.cea.nic.in/power_sec_reports/executive_summary/2010_03/1-2.pdf Accessed September 2010.

[41] MoP. "Annual Report, 2007-2008." July 2008; p. 109. http://www.cea.nic.in/about us/Annual%20Report/2007-08/annual report 07 08.pdf Accessed October 2010

[42] CEA. "Monthly Review of Power Sector (Executive Summary)" March 2010. http://www.cea.nic.in/power sec reports/executive summary/2010 03/27-33.pdf. Accessed October 2010.

[43] India's Five-Year Plans are planning documents for India's economy. See Section 2.3.1 of this report for a description of the planning process and renewable energy's role.

[44] CEA. "Summary Statement." http://www.cea.nic.in/planning/Capacity%20addition%20target%20during%2011th%20plan%20set%20by%20Planning%20Commission%20(Revised)-summary%20region%20wise.pdf. Accessed July 2010. "Summary of Likely Capacity Addition During 11th Plan." http://www.cea.nic.in/planning/Feasible% 20capacity% 20addition%20during%2011th%20Plan.pdf. Accessed July 2010.

[45] CEA. "Monthly Report: Executive Summary." March 2010; p. 11. http://www.cea.nic.in/power_sec_reports/executive_summary/2010_03/11.pdf. Accessed September 2010.

[46] CEA. "Monthly Report: Executive Summary." March 2010; p. 4. http://www.cea.nic.in/power_sec_reports/executive_summary/2010_03/4.pdf. Accessed September 2010.

[47] CEA. "Month ly Report: Executive Summary." March 2010.http://www.cea.nic.in/power sec reports/executive summary/2010 03/4.pdf. Accessed September 2010.

[48] Without inter alia deduction of own consumption of power stations (approximately 7%) and grid losses (approximately 3 0%).

[49] CEA. "All India Electricity Statistics—General Review." New Delhi: CEA. 2009; p. 6.

[50] CEA. "All India Electricity Statistics—General Review." New Delhi: CEA. 2009; p. 6.

[51] Coal India Limited. "CIL to Spend Rs 2,000 cr to Build 20 New Washeries." http://www.coalindia.in/NewsDisplay.aspx?NewsID=35&NewsType=2. Accessed July 2010.

[52] CEA and National Thermal Power Corporation. "Energy Efficiency in Thermal Power Generation." Second Meeting of the Indo-German Energy Forum. December 2007.

[53] Srivastava, N.K.; Ramesh, M.S.; Ramakrishnan, N.A. "NTPC R&M Case Study." http://docs.google. 5- 1 6May2008/1 5-05-08Presentations/II-NKSrivastava.pdf&hl=de&pid=bl&srcid=ADGEESjtsxI_JqXlS3VKwdXdUmADswmYTVHElM89fRStUfU B 9bpHnYs0JgCS8HWUjc7Nez9FjQTdn1WB8ola-Zsqrcv_1 oMl58k8 1 qu47PuGSgR-gtQp7hHYxa9LPgZKbDSuLT8CuWf&sig=AHIEtbRSI2CZQtPBCDxXXu1kDsbRqL1nzw. Accessed October 2010.

[54] Bhushan, C. "Challenge of the New Balance." New Delhi: Centre for Science and Environment, 2010; p. 25.

[55] PowerLine. "Industrial Tariff Trends." August 2009.

[56] PowerLine. "Impetus to T&D." June 2009.

[57] McKinsey. "Powering India: The Road to 2017." 2009 (unpublished).

[58] MNRE. "New & Renewable Energy." http://www.mnre.gov.in/achievements.htm. Accessed October 2010.

[59] CEA. "Monthly Review of Power Sector (Executive Summary)." September 2010. http://www.cea.nic.in/power_sec_reports/executive_summary/2010_09/1-2.pdf. Accessed October 2010.

[60] MNRE. "New & Renewable Energy." http://www.mnre.gov.in/achievements.htm. Accessed October 2010.

[61] MNRE. "New & Renewable Energy." http://www.mnre.gov.in/achievements.htm. Accessed October 2010.

[62] MNRE. "Jawaharlal Nehru National Solar Mission: Towards Building Solar India." http://mnre.gov.in/pdf/mission. Accessed October 2010.

[63] CSP Today. "India´s 'New Solar Mission' Opens Doors for Fast Development for the Concentrated Solar Thermal Industry Worldwide." 28 May 2010.

[64] Government of India Planning Commission. "Eleventh Five Year Plan—2007-12." p. 387. http://planningcommission.nic.in/plans/planrel/fiveyr/11th/11_v3/11th_vol3.pdf. Accessed September 2010.

[65] Government of India. "Jawaharlal Nehru National Solar Mission." http://mnre.gov.in/pdf/mission-. Accessed September 2010.

[66] Over a period of 25 years.

[67] MNRE. *Annual Report 2009-10*. http://mnre.gov.in/annualreport/2009-10EN/index.htm. Accessed May 2010.

[68] Muni Seva Ashram. http://munisevaashram.org/news6.html. Accessed July 2010.

[69] Economic Times. "1,000 Solar Rickshaws to Ferry Commonwealth Games Athletes." 21 February 2010. http://economictimes.indiatimes.com/news/news-by-industry/. Accessed October 2010.

[70] Economic Times. "Comonwealth Games 2010 to be 'Solar Powered' by Reliance." 7 April 2010. http://economictimes.indiatimes.com/news/news-by-industry/. Accessed July 2010.

[71] MNRE. http://www.mnre.gov.in/. Accessed June 21, 2010.

[72] DIREC 2010. "About the MNRE." http://direc2010.gov.in/about.html. Accessed June 21, 2010.

[73] Government of India Planning Commission. "Eleventh Five Year Plan 2007-2012: Agriculture, Rural Development, Industry, Services, and Physical Infrastructure." 2008; p. 385. http://planningcommission.nic.in/plans/planrel/fiveyr/11th/11 v3/11th vol3.pdf. Accessed April 2010.

[74] IREDA. "About IREDA—Background." http://www.ireda.gov.in/homepage1.asp?parent_category=1&category=6. Accessed September 2010.

[75] MoP. http://www.powermin.nic.in//. Accessed July 2010.

[76] CEA. http://www.cea.nic.in//. Accessed September 2010.

[77] CERC. http://cercind.gov.in//. Accessed September 2010.

[78] PFC. http://www.pfc.gov.in//. Accessed September 2010.

[79] PTC India Ltd. http://www.ptcindia.com//. Accessed September 2010.

[80] Rural Electrification Corporation Ltd. http://www.recindia.nic.in//. Accessed September 2010.

[81] Government of India Planning Commission. "Eleventh Five Year Plan 2007-2012: Agriculture, Rural Development, Industry, Services, and Physical Infrastructure." 2008. http://planningcommission.nic.in/plans/planrel/fiveyr/11th/11_v3/11th_vol3.pdf. Accessed April 2010.

[82] MoP. "The Electricity Act, 2003." http://www.powermin.nic.in/acts notification/electricity act2003/pdf/The%20Electricity%20Act 2003.pdf. Accessed October 2010.

[83] Baker & McKenzie. (2008). "Identifying Optimal Legal Frameworks for Renewable Energy in India." Renewable Energy and International Law. Asia-Pacific Partnership. p. 13. http://www.asiapacificpartnership. org/pdf/REDG-06-09%2520Final%2520ReportIndia.pdf&sa=U&ei=BgC3TITPBcXBswaPoZGPCQ&ved= 0CBMQFjAA&usg= AFQjCNGkQ6XCHcmHJkEc5rV5MIl384uEaw. Accessed October 2010.

[84] Government of India MoP National Electricity Policy. "The Gazette of India: Extraordinary Part I - Section 1." http://www.powermin.nic.in/whats new/national electricity policy.htm. Accessed September 2010.

[85] Government of India MoP. "Tariff Policy." http://www.powermin.nic.in/whats new/pdf/Tariff Policy.pdf. Accessed May 2010.

[86] Government of India Planning Commission. "Eleventh Five Year Plan 2007-2012: Agriculture, Rural Development, Industry, Services, and Physical Infrastructure." 2008. http://planningcommission.nic.in/plans/planrel/fiveyr/11th/11_v3/11th_vol3.pdf. Accessed April 2010.

[87] REEEP. "Policy and Regulatory Review—Special Report on India and Indian States." Renewable Energy & Energy Efficiency Partnership, 2009. http://www.reeep.org/file upload/296 tmpphpdxs0Zs.pdf. Accessed October 2010.

[88] Gtz. "Energiemarkt Indien 2010." p. 48. http://www.sciencerepository.org/in documents/IN1096.pdf. Accessed October 2010.

[89] Baker & McKenzie. "Identifying Optimal Legal Frameworks for Renewable Energy in India." Renewable Energy and International Law. Asia-Pacific Partnership, 2008. http://www.asiapacificpartnership.org/pdf/REDG-06-09%2520Final%2520Report_India_.pdf&sa=U&ei=BgC3TITPBcXBswaPoZGPCQ&ved=0CBMQFjAA&us g= AFQjCNGkQ6XCHcmHJkEc5rV5MIl384uEaw. Accessed October 2010.

[90] REEEP. "Policy and Regulatory Review—Special Report on India and Indian States." http://www.reeep.org/file upload/296 tmpphpdxs0Zs.pdf. Accessed October 2010.

[91] CEA. "Growth of Electricity Sector in India from 1947-2009." New Delhi: CEA, 2009.; pp. 3-4.

[92] MoP. "Overview." http://www.powermin.nic.in/transmission overview.htm. Accessed September 2010. "Power for All by 2012." http://www.powermin.nic.in/indian electricity scenario/power for all target.htm. Accessed September 2010

[93] Electricity Act 2003. Section 86(e). http://www.sciencerepository.org/in documents/IN1096.pdf. Accessed October 2010.

[94] Shah, S.R. "Feed-in Tariff & Renewable Purchase Obligation for Distribution Companies." Presented to the Associated Chambers of Commerce and Industry of India at the Conference on Mainstreaming Green

Energy— Wind, Biomass & Solar, Chennai. http://www.assocham.org/events/recent/showevent.php?id=426. April 29, 2010.

[95] CERC. "CERC Announces Renewable Energy Certificate (REC) Regulation—A Step Forward for Green Energy Promotion." Press Release, 18 January 2010.

[96] CERC. "REC Amendment Regulation." http://www.cercind.gov.in/Regulations/REC Amemdment Regulation 29 9 201 0.pdf. Accessed October 2010.

[97] The megawatts used throughout this chapter to represent wind power capacity represent wind turbine nameplate capacities. This is a convenient way to represent wind turbine sizes and the capacity of wind installations.

[98] World Wind Energy Association. "World Wind Energy Report 2009." http://www.wwindea.org/home/images s.pdf. Accessed October 2010.

[99] MNRE. "Achievements in New and Renewable Energy." http://mnre.gov.in. Accessed August 2010.

[100] Suzlon. http://www.suzlon.com. Accessed August 2010.

[101] Indian Wind Energy Association (INWEA). "Wind Energy Programme in India." http://www.inwea.org/aboutwindenergy.htm. Accessed August 2010.

[102] MNRE. "Wind Power Potential in India." http://mnre.gov.in/wpp.htm. Accessed August 2010.

[103] C-WET. "WPD Map." http://www.cwet.tn.nic.in/html/departments_wpdmap.html. Accessed August 2010.

[104] Indian Wind Turbine Manufacturers Association. "Indian Wind Energy and Economy." http://www.indianwindpower.com/iw energy economy.php. Accessed August 2010.

[105] GWEC. "Indian Wind Energy Outlook 2009." http://www.indianwindpower.com/pdf/GWEO A4 2008 India LowRes.pdf. Accessed August 2010.

[106] GWEC. *Global Wind Report 2008.* http://www.gwec.net/fileadmin/documents/Global%20Wind%202008%20Report.pdf. Accessed August 2010.

[107] Indian Wind Energy Association. http://www.inwea.org/. Accessed August 2010.

[108] MNRE. *Annual Report 2009-10.* http://mnre.gov.in/annualreport/2009-10EN/index.htm. Accessed August 2010.

[109] MNRE. "Achievements in New and Renewable Energy." http://mnre.gov.in. Accessed September 2010.

[110] MNRE. *Annual Report 2008-09*, http://mnre.gov.in/annualreport/2008-09EN/index.htm. Accessed August 2010.

[111] MNRE. *Annual Report 2009-10*. http://mnre.gov.in/annualreport/2009-10EN/index.htm. Accessed August 2010.

[112] IEA. "Electricity/Heat in India in 2007." http://www.iea.org/stats/electricitydata.asp?COUNTRY CODE=IN. Accessed August 2010.

[113] The capacity factors for China, Germany, Spain, and the United States were calculated using IEA's estimated wind power generation for 2007 (http://www.iea.org/stats/prodresult.asp?PRODUCT=Renewables; accessed August 2010), and the 2007 installed capacity values from GWEC's Global Wind 2007 report (http://www.gwec.net/index.php?id=90; accessed August 2010). GWEC's installed capacity value for India in 2007 was 8,000 MW; using this figure yields an estimated capacity factor of 16.6%.

[114] MNRE. "Unleashing the Wind Tiger." *Renewable Energy World;* May 2010.

[115] MNRE. "Generation Based Incentives (GBI) for Wind Power Projects." http://www.mnre.gov.in/gbischeme.htm. Accessed August 2010.

[116] MNRE. "Unleashing the Wind Tiger." *Renewable Energy World;* May 2010.

[117] CERC. "Tariff Order." http://www.cercind.gov.in/2009/November09/284-2009 final 3rdDecember09.pdf. Accessed September 2010.

[118] Shah, S.R. "Feed-in Tariff & Renewable Purchase Obligation for Distribution Companies." Presented to the Associated Chambers of Commerce and Industry of India at the Conference on Mainstreaming Green Energy— Wind, Biomass & Solar, Chennai. http://www.assocham.org/events/recent/showevent.php?id=426. April 29, 2010.

[119] MNRE. "Unleashing the Wind Tiger." *Renewable Energy World;* May 2010.

[120] IREDA. "Central Incentives." http://www.ireda.in/incentives.asp. Accessed August 2010.

[121] MNRE. "Generation Based Incentives (GBI) for Wind Power Projects." http://www.mnre.gov.in/gbischeme.htm. Accessed August 2010.

[122] Shah, S.R. "Feed-in Tariff & Renewable Purchase Obligation for Distribution Companies." Presented to the Associated Chambers of Commerce and Industry of India at the Conference on Mainstreaming Green Energy— Wind, Biomass & Solar, Chennai. http://www.assocham.org/events/recent/showevent.php?id=426. April 29, 2010.

[123] IREDA. "Central Incentives." http://www.ireda.in/incentives.asp. Accessed August 2010.

[124] MNRE. http://mnre.gov.in/gbi-scheme.htm. Accessed August 2010.

[125] REEEP. "Policy and Regulatory Review—Special Report on India and Indian States," 2010.

[126] IREDA. "Statewise List of Projects with WTGs Commissioned after 17.12.2009 Registered with IREDA for Unique Identification Numbers (UINs) under Generation Based Incentive (GBI) not Availing AD Under Section 44AB of Income Tax Act." http://www.ireda.gov.in/pdf/GBI%20Projects.pdf. Accessed October 18, 2010.

[127] Wiser, R.; Bolinger, M. "2009 Wind Technologies Market Report." LBNL-3716E. August 2010. http://eetd.lbl.gov/ea/emp/reports/lbnl-3716e.pdf. Accessed October 2010.

[128] Consolidated Energy Consultants. "Manufacturers-Wise Wind Electric Generators Installed in India," Wind Power India. http://www.windpowerindia.com/statmanuf.html. Accessed August 2010.

[129] MNRE. Presentation to Power-Gen India. 3 March 2010, p. 12.

[130] MNRE. Presentation to Power-Gen India. 3 March 2010, p. 10.

[131] GWEC. *Indian Wind Energy Outlook 2009.* http://www.indianwindpower.com/pdf/GWEO_A4_2008_India. Accessed August 2010.

[132] Suzlon. "RE Power-built 'Alpha Ventus' Starts Operations; Germany's First Offshore Wind Farm." Press Release. 28 April 2010. http://www.suzlon.com/media center/press release.aspx?l1=7&l2=3 1. Accessed September 2010.

[133] MNRE. "Unleashing the Wind Tiger." *Renewable Energy World;* May 2010.

[134] India Energy Portal. http://www.indiaenergyportal.org/subthemes link.php?text=wind&themeid=3. Accessed August 2010.

[135] Suzlon. "Key Financial Data." http://www.suzlon.com/investors. Accessed August 2010.

[136] Consolidated Energy Consultants. "Manufacturers-Wise Wind Electric Generators Installed in India." Wind Power India. http://www.windpowerindia.com/statmanuf.html. Accessed August 2010.

[137] "Saving Suzlon," *Forbes India.* 5 June 2009. http://business.in.com/article/cross-border/saving-suzlon/302/0. Accessed August 2010.

[138] Suzlon. "Unaudited Consolidated Financial Results for the Quarter Ended December 31, 2009." http://www.suzlon.com/images.

[139] Suzlon. Press Release, 29 May 2010. http://www.suzlon.com/media center/press release.aspx?l 1=7&l2=3 1. Accessed August 2010.

[140] Narasimhan, T.E. "BHEL Plans to Re-enter Wind Turbine Manufacturing Business." *Business Standard*, April 2010. http://www.business-standard.com/india/news/bhel-plans-to-re-enter-wind-turbine-manufacturing-business/390978/. Accessed August 2010.

[141] TATA Power. "Analyst Meet." Presented at the TATA Power Company Ltd Analyst Meet, March 2010. http://www.tatapower.com/investor-relations/pdf/analyst-presentation-mar-10.pdf. Accessed August 2010.

[142] "Green Energy: Airvoice Lines Up 13,000 MW." *Hindustan Times.* New Delhi, 5 February 2010. http://www.hindustantimes.com/News-Feed/markets/Green-energy 505712.aspx. Accessed October 2010.

[143] NHPC Limited. "Press Release." http://www.nhpcindia.com/english/scripts. Accessed August 2010.

[144] Neyveli Lignite Corporation Limited. "Press Release." http://www.nlcindia.com/investor/press release eng 270510.pdf. Accessed August 2010.

[145] Oil and Natural Gas Ltd. "ONGC's First Wind-farm, Having World's Highest Turbine Capacity, Inaugurated." http://www.ongcindia.com/press release 1 new.asp?fold=press&file=press348.txt. Accessed August 2010.

[146] Prashant, S. "ONGC Eyes 2,000 MW Wind Energy." Rediff India Abroad. 5 Februrary 2008. http://www.rediff.com/money/2008/feb/05wind.htm. Accessed October 18, 2010.

[147] GWEC. "Indian Wind Energy Outlook 2009." http://www.indianwindpower.com/pdf/GWEO A4 2008 India LowRes.pdf. Accessed August 2010.

[148] Gamesa. "Gamesa Reinforces its Presence in Asia by Starting Up its First Production Center in India." http://www.gamesacorp.com/en/press/press-releases/gamesa-reinforces-its-presence-in-asia-by-starting-up-its-first-production-center-in-india. Accessed August 2010.

[149] C-WET. *Pavan*; Iss. 16, January-March 2008; p. 7, www.cwet.tn.nic.in/Docu/news%20letter/english/Issue% 2016.pdf. Accessed August 2010.

[150] Energy Alternatives India. "India Offshore Wind—Status, Trends, Potential." http://eai.in/blog/2010/04/india-offshore-wind-status-trends-potential.html. Accessed October 2010.

[151] "Areva, Siemens, and GE Explore Offshore Wind-ow in India." *Economic Times.* http://economictimes. indiatimes.com/news/news-by-industry/ 1229.cms. Accessed August 2010.

[152] Chapter 4 discusses solar PV, CSP, and SWH. Solar process heat for industry and solar cooling are discussed briefly in Chapters 7 and 10, respectively.

[153] India Energy Portal. http://www.indiaenergyportal.org. Accessed August 2010.

[154] The megawatts used throughout this chapter to represent solar power capacity are peak megawatts, meaning the maximum rated power based on the equipment's power ratings under standard conditions.

[155] MNRE. "Jawaharlal Nehru National Solar Mission: Towards Building SOLAR INDIA." http://mnre.gov.in/pdf/mission. Accessed August 2010.

[156] MNRE. "Guidelines for New Grid Connected Solar Power Projects." July 2010. http://www.mnre.gov.in/pdf/ jnnsm-gridconnected-25072010.pdf. Accessed October 2010.

[157] MNRE. "Achievements." http://www.mnre.gov.in/achievements.htm. Accessed September 2010.

[158] MNRE. "Jawaharlal Nehru National Solar Mission: Towards Building SOLAR INDIA." http://mnre.gov.in/pdf/mission. Accessed August 2010.

[159] SWERA Renewable Energy Resource EXplorer. http://na.unep.net/swera _ims/map2/. Accessed August 2010.

[160] NREL. http://www.nrel.gov/international/ra india.html. Accessed September 2010.

[161] MNRE. http://mnre.gov.in/achievements.htm. Accessed September 2010.

[162] MNRE. *2009 Annual Report.* Table 1.1. http://mnre.gov.in/annualreport/2009-10EN/index.htm. Accessed September 2010.

[163] MNRE. http://mnre.gov.in/achievements.htm. Accessed September 2010

[164] Government of India Planning Commission. "Eleventh Five Year Plan 2007-12: Volume III Agriculture, Rural Development, Industry, Services, and Physical Infrastructure." p. 414, http://planningcommission.nic.in/plans/planrel/fiveyr/11th/11 v3/11th vol3.pdf. Accessed August 2010.

[165] Government of India Planning Commission. "Eleventh Five Year Plan 2007-12: Volume III Agriculture, Rural Development, Industry, Services, and Physical Infrastructure." p. 414, http://planningcommission.nic.in/plans/planrel/fiveyr/11th/11 v3/11th vol3.pdf. Accessed August 2010.

[166] MNRE. "Jawaharlal Nehru National Solar Mission: Towards Building SOLAR INDIA." http://mnre.gov.in/pdf/mission. Accessed August 2010.

[167] Mishra, R. "Solar Power: Govt Hopes to Seal Buy-Deals for 800 MW this Fiscal." *Hindu Business Line;* May 2010. http://www.thehindubusinessline.com/2010/05/11/stories/2010051153030400.htm. Accessed August 2010.

[168] CERC, New Delhi. Tariff Order. Petition No.53/2010. http://cercind.gov.in/2010/ORDER/February2010/53-2010 Suo-Motu RE Tariff Order FY20 10-11 .pdf. Accessed August 2010.

[169] MNRE. National Solar Mission, Guidelines for New Grid Connected Solar Power Projects, July 2010. http://www.mnre.gov.in/pdf/jnnsm-gridconnected-25072010.pdf. Accessed October 2010.

[170] MNRE. "Guidelines for Migration of Existing Under Development Grid Connected Solar Projects from Existing Arrangements to the Jawaharlal Nehru National Solar Mission (JNNSM)." http://www.mnre.gov.in/pdf/migration.

[171] MNRE. "Jawaharlal Nehru National Solar Mission: Towards Building SOLAR INDIA." http://www.mnre.gov.in/pdf/mission. Accessed August 2010.

[172] MNRE. National Solar Mission, Guidelines for New Grid Connected Solar Power Projects, July 2010. http://www.mnre.gov.in/pdf/jnnsm-gridconnected-25072010.pdf. Accessed October 2010.

[173] MNRE. National Solar Mission, Guidelines for New Grid Connected Solar Power Projects, July 2010. http://www.mnre.gov.in/pdf/jnnsm-gridconnected-25072010.pdf. Accessed October 2010.

[174] The net worth calculations: paid-up share capital plus reserves minus revaluation reserves minus intangible assets minus miscellaneous expenditures to the extent not written off or carry forward losses.

[175] MNRE. National Solar Mission, Guidelines for New Grid Connected Solar Power Projects, July 2010. http://www.mnre.gov.in/pdf/jnnsm-gridconnected-25072010.pdf. Accessed October 2010.

[176] MNRE. National Solar Mission, Guidelines for New Grid Connected Solar Power Projects, July 2010. http://www.mnre.gov.in/pdf/jnnsm-gridconnected-25072010.pdf. Accessed October 2010.

[177] Kumar, S. "Indian Solar Market Update." Solar Snippet. Credit Suisse. 23 June 2010.

[178] Kumar, S. "Indian Solar Market Update." Solar Snippet. Credit Suisse. 23 June 2010.

[179] Dutta, A.P. "New Gold Rush." *Down to Earth Magazine*; 1-5 June 2010.

[180] Kumar, S. "Indian Solar Market Update." Solar Snippet. Credit Suisse. 23 June 2010.

[181] Titan Energy. "Titan Energy Systems and Enfinity To Develop, Finance And Construct 1 GWp of PV Installations In Andhra Pradesh, India." 9 November 2009, http://www.titansolar.com/titan/press9nov2009.html. Accessed August 2010.

[182] Frost and Sullivan. "India Solar Photovoltaic Market Shines Bright as the Government Encourages Private Investments, finds Frost & Sullivan." 9 February 2010, http://www.frost. 92215618. Accessed October 2010.

[183] Frost and Sullivan. "India Solar Photovoltaic Market Shines Bright as the Government Encourages Private Investments, finds Frost & Sullivan." 9 February 2010, http://www.frost. 92215618. Accessed August 2010. Accessed October 2010.

[184] Frost and Sullivan. "Future of the Indian Solar PV Industry." 25 January 2010. http://www.frost. Accessed August 2010.

[185] MNRE. "Guidelines for Migration of Existing Under Development Grid Connected Solar Projects from Existing Arrangements to the Jawaharlal Nehru National Solar Mission (JNNSM)." http://www.mnre.gov.in/pdf/migration. Accessed August 2010.

[186] PLG Ltd. http://www.plgpower.com/about-milestones.asp. Accessed August 2010.

[187] TATA BP Solar. "Tata BP Solar Raises $78 Million for Further Investments in Solar Energy." 17 March 2008. http://www.tatabpsolar.com/future_invest.html. Accessed August 2010.

[188] TATA BP Solar. "Tata BP Solar Expands Solar Manufacturing Capacity by 62% to Serve Growing Solar Market in India." 28 April 2010. http://www.tatabpsolar.com/solar-expands.html. Accessed October 2010.

[189] Hindu Business Line. "Solar Power Scheme Draws Rs. 2.2-lakh cr Proposals." 14 April 2010. http://www.thehindubusinessline.com/2010/04/15/stories/2010041556760100.htm. Accessed August 2010.

[190] Hindu Business Line. "Solar Power Scheme Draws Rs. 2.2-lakh cr Proposals." 14 April 2010. http://www.thehindubusinessline.com/2010/04/15/stories/2010041556760100.htm. Accessed August 2010.

[191] New Energy. "End of the Solar Eclipse." September 2009, Iss. 5; p. 82.

[192] MNRE. "Jawaharlal Nehru National Solar Mission: Towards Building SOLAR INDIA." http://mnre.gov.in/pdf/mission. Accessed August 2010.

[193] Government of India Planning Commission. "Eleventh Five Year Plan 2007-12: Volume III Agriculture, Rural Development, Industry, Services, and Physical Infrastructure." p. 417, http://planningcommission.nic.in/plans/planrel/fiveyr/11th/11 v3/1 1th vol3.pdf. Accessed August 2010.

[194] The Hindu Business Line. TERI Newswire, Vol. 16., No. 10, 16-31 May 2010; p. 16.

[195] Dutta A.P. "New Gold Rush." *Down to Earth Magazine* 1-5 June 2010.

[196] Signet Solar "Signet Solar Expands Manufacturing Capacity with India Plant." 17 March 2008. http://www.signetsolar.com/pdf/Signet Solar Global Press Release mar 17 08.pdf. Accessed August 2010.

[197] TF Solar Power. http://www.tfsolarpower.com/. Accessed August 2010.

[198] Poseidon Solar. "Poseidon Solar Announces its Capability to Reclaim CdTe Thinfilm Solar Cells." 3 December 2009. http://poseidonsolar.com/latest-news/8.html. Accessed August 2010.

[199] Dutta A.P. "New Gold Rush." *Down to Earth Magazine* 1-5 June 2010.

[200] Azure. http://www.azurepower.com. Accessed August 2010.

[201] Bondre, K. "Interview—India's Azure Power Plans 100 MW Solar Projects" *Reuters* 19 May 2010. http://in.reuters.com/article/idINBMB01063720100519. Accessed August 2010.

[202] RIL Solar Group. 7 April 2010, http://www.ril.com/downloads/pdf/PR07042010R.pdf. Accessed August 2010.

[203] Moser Baer. http://www.moserbaerpv.in/about-overview.asp?links=ab1. Accessed August 2010.

[204] Moser Baer. 24 June 2010. http://www.moserbaerpv.in/media. Accessed August 2010.

[205] REN 21. "Renewables 2010: Global Status Report." http://www.ren21.net/globalstatusreport/REN21 GSR 2010 full.pdf. Accessed August 2010.

[206] Joint Research Center. "PV Status Report 2009." http://re.jrc.ec.europa.eu/refsys/pdf/PV-Report2009.pdf. Accessed August 2010.

[207] Walet, L. "Foreign Companies Take a Shine to India's $70 Billion Solar Programme." *Mint* 18 June 2010. http://www.livemint.com/2010/06/17215210/Foreign-companies-take-shine-t.html?atype=tp. Accessed August 2010.

[208] Walet, L. "Foreign Companies Take a Shine to India's $70 Billion Solar Programme." *Mint* 18 June 2010. http://www.livemint.com/2010/06/17215210/Foreign-companies-take-shine-t.html?atype=tp. Accessed August 2010.

[209] Kumar, S. "Indian Solar Market Update." Solar Snippet. Credit Suisse. 23 June 2010.

[210] Kumar, S. "Indian Solar Market Update." Solar Snippet. Credit Suisse. 23 June 2010.

[211] Astonfield. "Astonfield and Belectric Team up to Realize 5MW Solar Power Plant in Rajasthan, India." 11 May 2010. http://www.astonfield.com/press/pr/Astonfield-PR-Belectric-May2010-FINAL.pdf. Accessed August 2010.

[212] Astonfield. "Astonfield and Belectric Team up to Realize 5 MW Solar Power Plant in Rajasthan, India." 11 May 2010. http://www.astonfield.com/press/pr/Astonfield-PR-Belectric-May2010-FINAL.pdf. Accessed August 2010.

[213] IEA. "Technology Roadmap: Concentrating Solar Power." http://www.iea.org/papers/2010/csp_roadmap.pdf. Accessed August 2010.

[214] IEA. Solar Paces, http://www.solarpaces.org/News/Projects/India. Accessed August 2010.

[215] DESERTEC Foundation. "Red Paper" p. 7. http://www.desertec.org/fileadmin/downloads/desertecfoundation redpaper 3rd-edition english.pdf. Accessed August 2010.

[216] SWERA Renewable Energy Resource EXplorer. http://na.unep.net/swera _ims/map2/. Accessed August 2010.

[217] NREL. http://www.nrel.gov/international/ra india.html. Accessed September 2010.

[218] Trieb, F.; Shillings, C.; O'Sullivan, M.; Pregger, T.; Hoyer-Klick, C. "2009: Global Potential of Concentrating Solar Power." SolarPaces Conference Berlin. http://www.dlr.de/tt/en/Portaldata/41/Resources/dokumente/institut/system/projects/reaccess/DNI-Atlas-SPBerlin 20090915-04-Final-Colour.pdf. Accessed October 2010.

[219] Adapted from Trieb, F.; Shillings, C.; O'Sullivan, M.; Pregger, T.; Hoyer-Klick, C. "2009: Global Potential of Concentrating Solar Power." SolarPaces Conference Berlin. http://www.dlr.de/tt/en/Portaldata/41/Resources/dokumente/institut/system/projects/reaccess/DNI-Atlas-SPBerlin 20090915-04-Final-Colour.pdf. Accessed October 2010.

[220] The Hindu Business Line. TERI Newswire, Vol. 16., No. 8, 16-31 April 2010; p. 8.

[221] The Hindu Business Line. TERI Newswire, Vol. 16., No. 10, 16-31 May 2010.

[222] ACME. "Solar Thermal Technology." http://www.acme.in/solar-thermal-technology. Accessed August 2010.

[223] NTPC. "Foray into Renewable Energy." https://www.ntpc.co.in/index.php?option=com_content&view=article&id=206&lang=en. Accessed August 2010.

[224] Gujarat Energy Development Agency. http://www.geda.org.in/pdf/solar allotment webnote.pdf. Accessed September 2010.

[225] Rajastan Renewable Energy Corporation Ltd. "Note on Solar Energy Power Plants in Rajasthan." 2010. www.rrecl.com/Note%20on%20Solar%20Energy.pdf. Accessed August 2010.

[226] Government of India Planning Commission. "Eleventh Five Year Plan 2007-12: Volume III Agriculture, Rural Development, Industry, Services, and Physical Infrastructure." p. 414, http://planningcommission.nic.in/plans/planrel/fiveyr/11th/11 v3/11th vol3.pdf. Accessed August 2010.

[227] REEEP. "Policy and Regulatory Review—Special Report on India and Indian States, 2010." http://www.reeep.org/file_upload/296_tmpphpdxs0Zs.pdf. Accessed September 2010.

[228] REEEP. "Policy and Regulatory Review—Special Report on India and Indian States, 2010." http://www.reeep.org/file_upload/296_tmpphpdxs0Zs.pdf. Accessed September 2010.

[229] MNRE. National Solar Mission. Guidelines for New Grid Connected Solar Power Projects, July 2010. http://www.mnre.gov.in/pdf/jnnsm-gridconnected-25072010.pdf. Accessed October 2010.

[230] CERC. Tariff Order. Petition No.53/2010. 26 February 2010. http://cercind.gov.in/2010/ORDER/February2010/53-2010_Suo-Motu_RE_Tariff_Order_FY2010-11.pdf. Accessed August 2010.

[231] MNRE. National Solar Mission. Guidelines for New Grid Connected Solar Power Projects, July 2010. http://www.mnre.gov.in/pdf/jnnsm-gridconnected-25072010.pdf. Accessed October 2010.

[232] MNRE. National Solar Mission. Guidelines for New Grid Connected Solar Power Projects, July 2010. http://www.mnre.gov.in/pdf/jnnsm-gridconnected-25072010.pdf. Accessed October 2010.

[233] MNRE. National Solar Mission. Guidelines for New Grid Connected Solar Power Projects, July 2010. http://www.mnre.gov.in/pdf/jnnsm-gridconnected-25072010.pdf. Accessed October 2010.

[234] MNRE. National Solar Mission. Guidelines for New Grid Connected Solar Power Projects, July 2010. http://www.mnre.gov.in/pdf/jnnsm-gridconnected-25072010.pdf. Accessed October 2010.

[235] MNRE. National Solar Mission. Guidelines for New Grid Connected Solar Power Projects, July 2010. http://www.mnre.gov.in/pdf/jnnsm-gridconnected-25072010.pdf. Accessed October 2010.

[236] Gujarat Energy Development Agency. Tariff Order, 29 January 2010, p. 29. http://www.gercin.org/docs/Orders/Nonconv%20orders/Year%202010/Order%202-2010.pdf. Accessed September 2010.

[237] MNRE. "Jawaharlal Nehru National Solar Mission: Towards Building SOLAR INDIA." p. 6. http://mnre.gov.in/pdf/mission. Accessed August 2010.

[238] MNRE. "Jawaharlal Nehru National Solar Mission: Towards Building SOLAR INDIA." p. 6–7. http://mnre.gov.in/pdf/mission. Accessed August 2010.

[239] DESERTEC Foundation. "Red Paper." http://www.desertec.org/fileadmin/downloads/desertec-foundation redpaper 3rd-edition english.pdf. Accessed August 2010.

[240] NREL. "U.S. Parabolic Trough Power Plant Data." http://www.nrel.gov/csp/troughnet/power_plant_data.html. Accessed August 2010.

[241] Solar Millennium. "The Construction of the Andasol Power Plants." http://www.solarmillennium.de/front content.php?idart=1 55&lang=2. Accessed August 2010.

[242] DESERTEC Foundation. http://www.desertec.org/en/concept/technologies/. Accessed August 2010.

[243] Walet, L. "Foreign Companies Take a Shine to India's $70 Billion Solar Programme." *Mint* 18 June 2010. http://www.livemint.com/2010/06/17215210/Foreign-companies-take-shine-t.html?atype=tp. Accessed August 2010.

[244] Siemens. "Siemens to Supply Solar Receivers for New Solar Power Plant in Spain." 24 March 2010, http://www.siemens.com/press/en/pressrelease/?press=/en/pressrelease/2010/renewable energy/ere201 003049.h tm. Accessed August 2010.

[245] Siemens. "Siemens to Supply Solar Receivers for New Solar Power Plant in Spain." 24 March 2010, http://www.siemens.com/press/en/pressrelease/?press=/en/pressrelease/2010/renewable_energy tm. Accessed August 2010.

[246] REN21. "Renewables 2010 Global Status Report." Paris: REN21 Secretariat. 2010. http://www.ren21.net/globalstatusreport/REN21 GSR 2010 full.pdf. Accessed October 2010.

[247] REN21. "Renewables 2010 Global Status Report." Paris: REN21 Secretariat. 2010. http://www.ren21.net/globalstatusreport/REN21 _GSR _2010 _full.pdf. Accessed October 2010.

[248] MNRE. "A Brief on Solar Water Heating Systems." http://www.mnre.gov.in/Solar-water-. Accessed October 2010.

[249] OPET-TERI & HECOPET. "Status of Solar Thermal Technologies and Markets in India and Europe—An OPET International Action 'Enhancement Of Market Penetration Of Solar Thermal Technologies.'" p. 5. http://www.teriin.org/opet/reports/solarthermal.pdf. Accessed October 2010.

[250] GKS. "Solar Water Heaters in India: Market Assessment Studies and Surveys for Different Sectors and Demand Segments." 20 January 2010. http://mnre.gov.in/pdf/greentech-SWH-MarketAssessment-report.pdf. Accessed August 2010.

[251] MNRE. *Annual Report 2002-03*. http://mnre.gov.in/annualreport/2002_2003_English/ch4_pg2.htm. Accessed October 2010.

[252] MNRE. *Annual Report 2008-09*. http://mnre.gov.in/annualreport/2008-09EN/renewable-energy-. Accessed October 2010.

[253] MNRE. *Annual Report 2009-10*. Table 1.1, No. 14. http://mnre.gov.in/annualreport/2009-1 0EN/Chapter1/chapter1 _1 .htm. Accessed August 2010.

[254] MNRE. *Annual Report 2009-10*. Table 1.1, No. 14. http://mnre.gov.in/annualreport/2009-1 0EN/Chapter1/chapter1 1 .htm. Accessed August 2010.

[255] MNRE. *Annual Report 2009-10*. http://mnre.gov.in/annualreport/2009- 1 0EN/Chapter%205/chapter%205 1 .htm. Accessed October 2010.

[256] MNRE. *Annual Report 2009-10*. http://mnre.gov.in/annualreport/2009- 1 0EN/Chapter%205/chapter%205 1.htm. Accessed October 2010.

[257] REN21. "Renewables 2010 Global Status Report." Paris: REN21 Secretariat. 2010. http://www.ren21.net/globalstatusreport/REN21 _GSR _2010 _full.pdf. Accessed October 2010.

[258] GKS. "Solar Water Heaters in India: Market Assessment Studies and Surveys for Different Sectors and Demand Segments." 20 January 2010. http://mnre.gov.in/pdf/greentech-SWH-MarketAssessment-report.pdf. Accessed August 2010.

[259] GKS. "Solar Water Heaters in India: Market Assessment Studies and Surveys for Different Sectors and Demand Segments." 20 January 2010. http://mnre.gov.in/pdf/greentech-SWH-MarketAssessment-report.pdf. Accessed August 2010. Note this report assumes 85% of installed collector area was functioning at the time of their investigation, which equates to a total functional installed collector area of 3.1 million m^2, which differs from the MNRE official estimate of total installed collector area through 2009 of 3.4 million m^2.

[260] Greentech Knowledge Solutions Ltd. "Solar Water Heaters in India: Market Assessment Studies and Surveys for Different Sectors and Demand Segments." 20 January 2010. http://mnre.gov.in/pdf/greentech-SWH-MarketAssessment-report.pdf. Accessed August 2010.

[261] Greentech Knowledge Solutions Ltd. "Solar Water Heaters in India: Market Assessment Studies and Surveys for Different Sectors and Demand Segments." 20 January 2010. http://mnre.gov.in/pdf/greentech-SWH-MarketAssessment-report.pdf. Accessed August 2010.

[262] Greentech Knowledge Solutions Ltd. "Solar Water Heaters in India: Market Assessment Studies and Surveys for Different Sectors and Demand Segments." 20 January 2010. http://mnre.gov.in/pdf/greentech-SWH-MarketAssessment-report.pdf. Accessed August 2010.

[263] Greentech Knowledge Solutions Ltd. "Solar Water Heaters in India: Market Assessment Studies and Surveys for Different Sectors and Demand Segments." 20 January 2010. p. 35. http://mnre.gov.in/pdf/greentech-SWH-MarketAssessment-report.pdf. Accessed August 2010.

[264] UNDP. *Global Solar Water Heating Market Transformation and Strengthening Initiative.* 2009. http://www.undp.org.in/sites/default/files/Global-Solar-Water-Heating.pdf. Accessed October 2010.

[265] GRIHA, an acronym for Green Rating for Integrated Habitat Assessment, is the national rating system of India. This green building rating system has been developed by TERI, after a thorough study and understanding the current internationally accepted green building rating systems and the prevailing building practices in India. The GRIHA rating system consists of 34 criteria categorized under various sections such as Site Selection and Site Planning, Building Operation and Maintenance, and Innovation. For more information, see http://www.grihaindia.org/. Accessed August 2010.

[266] MNRE. *Annual Report 2009-10*. http://mnre.gov.in/annualreport/2009-10EN/index.htm. Accessed July 2010.

[267] MNRE. "Achievements as of March 31, 2010." http://www.mnre.gov.in/achievements.htm. Accessed July 2010.

[268] Palsh, O. "Small Hydro Power: Technology and Current Status." *Renewable Sustainable Energy Rev.*; Vol. 6, 2002; pp. 537-556.

[269] MNRE. *Annual Report 2009-10*. http://mnre.gov.in/annualreport/2009-10EN/index.htm. Accessed July 2010.

[270] MNRE. "Achievements as of March 31, 2010." http://www.mnre.gov.in/achievements.htm. Accessed July 2010.

[271] MNRE. *Annual Report 2009-10*. http://mnre.gov.in/annualreport/2009-10EN/index.htm. Accessed October 2010.

[272] MNRE. *Annual Report 2009-10*. http://mnre.gov.in/annualreport/2009-10EN/index.htm. Accessed July 2010.

[273] Government of India Planning Commission. "Eleventh Five Year Plan 2007-12: Agriculture, Rural Development, Industry, Services, and Physical Infrastructure." New Delhi, India: Oxford University Press, 2008; p. 414. http://planningcommission.nic.in/plans/planrel/fiveyr/11th/11 v3/1 1th vol3.pdf. Accessed July 2010.

[274] Government of India Planning Commission. "Eleventh Five Year Plan 2007-12: Agriculture, Rural Development, Industry, Services, and Physical Infrastructure." New Delhi, India: Oxford University Press, 2008; p. 414. http://planningcommission.nic.in/plans/planrel/fiveyr/11th/11 v3/1 1th vol3.pdf. Accessed July 2010.

[275] MNRE. *Annual Report 2009-10*. http://mnre.gov.in/annualreport/2009-10EN/index.htm. Accessed July 2010.

[276] MNRE. "Achievements as of March 31, 2010." http://www.mnre.gov.in/achievements.htm. Accessed July 2010.

[277] Kesharwani, M.K. "Overview of Small Hydro Power Development in Himalayan Region." Himalayan Small Hydropower Summit; 12-13 October 2006 in Dehradun; p. 51. http://ahec.org.in/acads/HSHS/Presentations/Links/Technical%20Papers/Overview%20of%20SHP%20Develo p ment/Mr%20MK%20Kesharwani _Overview%20of%20SHP%20Development.pdf. Accessed September 2010.

[278] MNRE. "Small Hydro Power Programme." http://mnre.gov.in/adm-approvals/shp scheme.pdf. Accessed July 2010.

[279] MNRE. *Annual Report 2009-10*. http://mnre.gov.in/annualreport/2009-10EN/index.htm. Accessed July 2010.

[280] MNRE. *Annual Report 2009-10*. http://mnre.gov.in/annualreport/2009-10EN/index.htm. Accessed July 2010.

[281] MNRE. *Annual Report 2009-10*. http://mnre.gov.in/annualreport/2009-10EN/index.htm. Accessed July 2010.

[282] MNRE. "Small Hydro Power Programme." http://www.mnre.gov.in/prog-smallhydro.htm. Accessed October 2010.

[283] CERC. "Regulations." http://cercind.gov.in/Regulations/Final SOR RE Tariff Regulations to upload 7 oct 09.pdf. Accessed September 2010.

[284] CERC. "Regulations." http://cercind.gov.in/Regulations/Final SOR RE Tariff Regulations to upload 7 oct 09.pdf. Accessed September 2010.

[285] CERC. "Tariff Order." http://www.cercind.gov.in/2009/November09/284-2009_final_3rdDecember09.pdf. Accessed September 2010.

[286] MNRE. "Small Hydro Power Programme." http://mnre.gov.in/prog-smallhydro.htm. Accessed September 2010.

[287] MNRE. *Annual Report 2009-10*. p. 54. http://mnre.gov.in/annualreport/2009-1 0EN/Chapter%206/chapter%206 1 .htm. Accessed July 2010.

[288] MNRE. *Annual Report 2009-10*. p. 54. http://mnre.gov.in/annualreport/2009-1 0EN/Chapter%206/chapter%206 1 .htm. Accessed July 2010.

[289] MNRE. http://mnre.gov.in. Accessed July 2010.

[290] MNRE. "Renewable Energy in North Eastern States." p. 86. http://mnre.gov.in/annualreport/2009-1 0EN/Chapter%209/chapter%209 1 .htm. Accessed July 2010.

[291] MNRE. "Renewable Energy in North Eastern States." p. 86. http://mnre.gov.in/annualreport/2009-10EN/Chapter%209/chapter%209 _1 .htm. Accessed July 2010.

[292] Ashkay Urja. "National Workshop on Village Energy Security Program." Vol. 2, Issue 16, June 2009; p. 43. http://www.mnre.gov.in/akshayurja/june09-e.pdf. Accessed October 2010.

[293] TRIPCO. "Particulars of the Organization, Functions and Duties." http://www.aptribes.gov.in/html/tripco/1.pdf. Accessed October 2010.

[294] Government of India Planning Commission. "Integrated Energy Policy: Report of the Expert Committee." 2006; p. 90. http://planningcommission.nic.in/reports/genrep/rep intengy.pdf. Accessed July 2010.

[295] MNRE. http://www.mnre.gov.in/shp-rand.htm. Accessed September 2010. Alternate Hydro Energy Centre IIT; Roorkee; MNRE. "Standards/Manuals/Guidelines for Small Hydro Development." http://www.iitr.ac.in/PageUploads/files/112%2520Guidelines%2520for%2520modernization%2520and%2520 renovation%2520of%2520SHP%2520stations.pdf. Accessed September 2010.

[296] Awasthi, S.R. "Development of Renewable Energy Technologies in India, The Role of BHEL" http://www.indiaenvironmentportal.org.in/files/BHEL.pdf June 2009. Accessed October 2010.

[297] Awasthi, S.R. "Development of Renewable Energy Technologies in India, The Role of BHEL" http://www.indiaenvironmentportal.org.in/files/BHEL.pdf June 2009. Accessed October 2010.

[298] Awasthi, S.R. "Development of Renewable Energy Technologies in India, The Role of BHEL" http://www.indiaenvironmentportal.org.in/files/BHEL.pdf June 2009. Accessed October 2010.

[299] Pentaflo Hydro Engineers. http://www.pentaflo.com/about.html. Accessed October 2010.

[300] Flovel Mecamidi. http://www.flovel.in/About-Us.html. Accessed October 2010.

[301] REN21. "Global Status Report 2010." http://www.ren21.net/globalstatusreport/g2010.asp. Accessed October 2010.

[302] NREL. "Renewable Energy Policy in China: Financial Incentives." http://www.nrel.gov/docs/fy04osti/36045.pdf. Accessed October 2010.

[303] Limin, H. "Financing Rural Renewable Energy: A Comparison Between China and India." *Renew. Sustainable Energy Rev.;* Vol. 13, 2009; pp. 1096-1103.

[304] UNIDO. http://www.unidorc.org/emc_shp.htm. Accessed October 2010.

[305] UNIDO. http://www.unidorc.org/emc_act.htm. Accessed July 2010.

[306] MNRE. "List of Equipment Manufacturers of Small Hydro Turbines." http://mnre.gov.in/manufacurerssht.htm Accessed October 2010.

[307] European Energy Forum. "Energy Policy in India Dinner Debate." http://www.europeanenergyforum.eu/archives/european-energy-. Accessed September 2010.

[308] Karandikar, V.; Rana, A. "Future of Energy Options for India in an Independent World." Reliance Industries Ltd.; p. 6. http://www.worldenergy.org/documents/p001145.pdf. Accessed September 2010.

[309] Ravindranath, N.H.; Balachandra, P. "Sustainable Bioenergy for India: Technical, Economic and Policy Analysis." *Energy J.;* Iss. 34, 2009; pp. 1003–101 3.

[310] Maithani, P.C. MNRE. Personal communication. October 2010.

[311] MNRE. "Annual Report 2009-10." http://www.mnre.gov.in/annualreport/2009- 1 0EN/Chapter%205/chapter%205 _1 .htm. Accessed October 2010.

[312] Maithani, P.C. MNRE. Personal communication. October 2010.

[313] MNRE. "Annual Report 2009-10." http://mnre.gov.in/annualreport/2009-10EN/index.htm. Accessed May 2010.

[314] Government of India Planning Commission. "Eleventh Five Year Plan 2007-12: Agriculture, Rural Development, Industry, Services, and Physical Infrastructure." http://planningcommission.nic.in/plans/planrel/fiveyr/11th/11 v3/1 1th vol3.pdf. Accessed April 2010.

[315] New Energy India. "RE Policy Overview." http://www.newenergyindia.org/Policy%20Page.htm. Accessed May 2010.

[316] MNRE. *Annual Report 2009-10*. http://mnre.gov.in/annualreport/2009-10EN/index.htm. Accessed May 2010.

[317] MNRE. *Annual Report 2009-10*. http://mnre.gov.in/annualreport/2009-10EN/index.htm. Accessed May 2010.

[318] MNRE. *Annual Report 2009-10*. http://mnre.gov.in/annualreport/2009-10EN/index.htm. Accessed May 2010.

[319] MNRE. *Annual Report 2009-10*. http://mnre.gov.in/annualreport/2009-10EN/index.htm. Accessed May 2010.

[320] Financial Express. (2010). "Haryana Looks at Biogas Plants to Boost Green Energy Use." 8 February 2010. http://www.financialexpress.com/news/haryana-looks-at-biogas. Accessed May 2010.

[321] "Länderprofil Indien." Berlin: Deutsche Energie-Agentur GmbH, 2007 (update February 2010).

[322] Developpp.de. http://www.developpp.de. Accessed July 16, 2010.

[323] MNRE. *Annual Report 2009-10*. http://mnre.gov.in/annualreport/2009-10EN/index.htm. Accessed May 2010.

[324] Ashden Awards. "Clean Cooking and Income Generation from Biogas Plants in Karnataka." http://www.ashdenawards.org/files/reports/SKG Sangha 2007 Technical report.pdf. Accessed July 2010.

[325] Ashden Awards. "Management of Domestic and Municipal Waste at Source Produces Biogas for Cooking and Electricity Generation." http://www.ashdenawards.org/files/reports/Biotech_India. Accessed July 2010.

[326] MNRE. "Biomass Resource Atlas of India." v2.0; 2009. Biomass data by state 2002-04, http://lab.cgpl.iisc.ernet.in/Atlas/Tables/Tables.aspx. Accessed October 2010.

[327] MNRE. "Biomass Resource Atlas of India." v2.0; 2009. Biomass data by state 2002-04, http://lab.cgpl.iisc.ernet.in/Atlas/Tables/Tables.aspx. Accessed October 2010.

[328] MNRE. *Annual Report 2008-2009*. http://www.mnre.gov.in/annualreport/2008-09EN/power-forrenewables.htm. Accessed October 2010.

[329] MNRE. *Annual Report 2009-2010*. http://www.mnre.gov.in/annualreport/2009- 10EN/Chapter%206/chapter%206 1 .htm. Accessed October 2010.

[330] MNRE. "Achievements." http://mnre.gov.in/achievements.htm. Accessed October 2010.

[331] MNRE. *Annual Report 2009-2010*. http://www.mnre.gov.in/annualreport/2009-10EN/Chapter%205/chapter%205 1.htm. Accessed October 2010.

[332] MNRE. "Achievements." http://mnre.gov.in/achievements.htm. Accessed October 2010.

[333] Government of India Planning Commission. "Eleventh Five Year Plan 2007-12: Agriculture, Rural Development, Industry, Services, and Physical Infrastructure." http://planningcommission.nic.in/plans/planrel/fiveyr/11th/11 v3/11th vol3.pdf. Accessed April 2010.

[334] Shah, S.R. "Feed-in Tariff and Renewable Purchase Obligation for Distribution Companies." Presentation to the Associated Chambers of Commerce and Industry of India at the Conference on Mainstreaming Green Energy—Wind, Biomass & Solar on April 29, 2010 in Chennai. http://www.assocham.org/events/recent/showevent.php?id=426. Accessed August 2010.

[335] IREDA. "Incentives Available in the Sector (Biomass Power Generation)." http://www.ireda.gov.in/homepage1.asp?parent_category=2&sub_category=22&category=27. Accessed May 2010.

[336] MNRE. *Annual Report 2009-10*. http://mnre.gov.in/annualreport/2009-10EN/index.htm. Accessed May 2010.

[337] New Energy India. "RE Policy Overview." http://www.newenergyindia.org/Policy%20Page.htm. Accessed May 2010.

[338] Naik, S.Y.; Yadav, S.S. "Renewable Energy Policy and Regulation Focusing on RPS." Presentation to the Associated Chambers of Commerce and Industry of India at the Conference on Mainstreaming Green Energy – Wind, Biomass & Solar on Apr. 29, 2010, in Chennai. http://www.assocham.org/events/recent/showevent.php?id=426. Accessed August 2010.

[339] IREDA. "Incentives Available in the Sector (Biomass Power Generation)." http://www.ireda.in/homepage1.asp?parent category=2&sub category=27&category=91. Accessed May 2010.

[340] Jayakumar, K.B. "A2Z Plans 500-MW Power Plant from Municipal Waste." *Business Standard* June 6, 2010. http://www.business-standard.com/india/news/a2z-plans-500-mw-power-plantmunicipalwaste/09/23/397203/. Accessed July 2010.

[341] Gera, K.A. "And Now, Power Generation from Poultry Litter!" *Business Standard* June 8, 2010. http://www.business-standard.com/india/news/and-now-power-generationpoultry-litter/397347/. Accessed July 2010.

[342] Ashden Awards. "Replacing Diesel Generators with Biomass Gasification Systems." http://www.ashdenawards.org/files/reports/Saran case study 2009 final map.pdf. Accessed July 2010.

[343] Ashden Awards. "School Cookstoves Running on Crop Waste in North India." http://www.ashdenawards.rg/files/reports/Nishant%202005%20Technical%20report.pdf. Accessed July 2010.

[344] "India Jatropha Electrification Initiative of Winrock International India." Good Practice Assessment Form. http://www.compete-bioafrica.net/bestpractice/COMPETE-032448-GoodPractice-CaseStudy1-India.pdf. Accessed July 2010.

[345] United States Department of Agriculture—Foreign Agricultural Service, Global Agricultural Information Network. "India Biofuels Annual 2009. http://gain.fas.usda.gov/Recent%20GAIN%20Publications/General%20Report_New%20Delhi_India_6-12- 2009.pdf. Accessed August 2010.

[346] Jog, S. "Ethanol Price Fixed at Rs 27 Per Litre for Blended Fuel." *Business Standard*. April 8, 2010. http://www.business-standard.com/india/news/ethanol. Accessed August 2010.

[347] Gonsalves, J. "An Assessment of the Biofuels Industry in India." United Nations Conference on Trade and Development, October 18, 2006, Geneva.

[348] United States Department of Agriculture—Foreign Agricultural Service, Global Agricultural Information Network. "India Biofuels Annual 2009. http://gain.fas.usda.gov/Recent%20GAIN%20Publications/ General%20Report New%20Delhi India 6-12- 2009.pdf. Accessed August 2010.

[349] Gonsalves, J. "An Assessment of the Biofuels Industry in India." United Nations Conference on Trade and Development, October 18, 2006, Geneva.

[350] MPNG. "Basic Statistics on Indian Petroleum and Natural Gas 2008-09." September 2009, New Delhi, India. Accessed October 2010.

[351] United States Department of Agriculture—Foreign Agricultural Service, Global Agricultural Information Network. "India Biofuels Annual 2009. http://gain.fas.usda.gov/Recent%20GAIN%20Publications/ General%20Report_New%20Delhi_India_6-12- 2009.pdf. Accessed August 2010.

[352] Gonsalves, J. "An Assessment of the Biofuels Industry in India." United Nations Conference on Trade and Development, 18 October 2006, Geneva.

[353] United States Department of Agriculture—Foreign Agricultural Service, Global Agricultural Information Network. "India Biofuels Annual 2009." http://gain.fas.usda.gov/Recent%20GAIN%20Publications/ General%20Report New%20Delhi India 6-12- 2009.pdf. Accessed August 2010.

[354] Punia, M.S. "Current Status of R&D on Jatropha for Sustainable Biofuels Production in India." Ministry of Agriculture. USDA Global Conference on Agricultural Biofuels, Minneapolis, MN, 20-22 August 2007.

[355] United States Department of Agriculture—Foreign Agricultural Service, Global Agricultural Information Network. "India Biofuels Annual 2009." http://gain.fas.usda.gov/Recent%20GAIN%20Publications/ General%20Report New%20Delhi India 6-12- 2009.pdf. Accessed August 2010.

[356] Singh, S. "India—Biofuels Annual." United States Department of Agriculture's Foreign Agricultural Service. GAIN Report No. IN9080; 2009.

[357] Government of India Planning Commission. "Eleventh Five Year Plan 2007-2012: Agriculture, Rural Development, Industry, Services, and Physical Infrastructure." http://planningcommission.nic.in/plans/planrel/ fiveyr/11th/11 v3/1 1th vol3.pdf. Accessed April 2010.

[358] Singh, S. "India – Biofuels Annual." United States Department of Agriculture's Foreign Agricultural Service. GAIN Report No. IN9080; 2009.

[359] MNRE. "National Policy on Biofuels." http://www.mnre.gov.in/policy. Accessed September 2010.

[360] Singh, S. "India—Biofuels Annual." United States Department of Agriculture's Foreign Agricultural Service. GAIN Report No. IN9080; 2009.

[361] Jatropha World. http://www.jatrophabiodiesel.org/indianScene.php. Accessed August 2010.

[362] Mint. "Biodiesel Producers Looking for Tie-ups with Telecom Tower Firms." 22 April 2010. http://www.livemint.com/2010/04/21233621/Biodiesel-producers. Accessed July 2010.

[363] The Pioneer. "Rlys Looks at Bio-fuel to Cut Costs." 5 July 2010. http://www.dailypioneer.com/267113/Rlys-looks-at-bio-fuel-to-cut-costs.html. Accessed July 2010.

[364] REN21. http://www.ren21 .net/pdf/REN21 LRE2009 Jun12.pdf. Accessed August 2010.

[365] IEA. "The Electricity Access Database." http://www.iea.org/weo/database electricity/electricity access database.htm. Accessed October 2010.

[366] IRADe and ENERGIA. "Gender Analyses of Renewable Energy in Inida – Present Status, Issues, Approaches and New Initiatives." 2009; p. 11. http://www.irade.org/Gender%20Cover.pdf. Accessed October 2010.

[367] Government of India. "Report of the Expert Group on a Viable and Sustainable System of Pricing of Petroleum Products." February 2010; p. 62. http://petroleum. Accessed October 2010

[368] Government of India. "Report of the Expert Group on a Viable and Sustainable System of Pricing of Petroleum Products." February 2010; p. 62. http://petroleum. Accessed October 2010 and http://www.bioenergylists.org/stovesdoc/Nariphaltan/housenergy.pdf. Accessed October 2010.

[369] MoP. "Historical Background of Legislative Initiatives." http://www.powermin.nic.in/indian electricity scenario/pdf/Historical%20Back%20Ground.pdf. Accessed September 2010.

[370] RGGVY. "Definition of Electrified Village." http:// rggvy.gov.in/rggvy/rggvyportal/def_elect_vill.htm. Accessed September 2010.

[371] MoP. "Rural Electrification Policy." p. 6. http://www.powermin.nic.in/whats new/pdf/RE%20Policy.pdf. Accessed August 2010.

[372] RGGVY. http://www.rggvy.gov.in/rggvy/rggvyportal/index.html. Accessed October 2010.

[373] MoP. RGGVY brochure. http://recindia.gov.in/download/rggvy brochure.pdf. Accessed October 2010.

[374] IEA; Niez, A. "Comparative Study on Rural Electification Policies in Emerging Economies – Keys to Successful Policies. March 2010; p. 67. http://www.iea.org/papers/2010/rural_elect.pdf. Accessed October 2010.

[375] MoP. "Guidelines for DDG." http://www.recindia.nic.in/download/DDG guidelines.pdf. Accessed October 2010.

[376] RGGVY. "Project Outlay, Coverage, and Progress So Far." http://www.rggvy.gov.in/rggvy/rggvyportal/plgsheet frame3.jsp. Accessed October 2010.

[377] MNRE. "Jawaharlal Nehru National Solar Mission. Toward building SOLAR INDIA." http://mnre.gov.in/pdf/mission. Accessed October 2010.

[378] MNRE. "Guidelines for Off-grid and Decentralized Application." http://www.mnre.gov.in/pdf/jnnsmg1 7061 0.pdf. Accessed October 2010.

[379] MNRE. "Details of Projects Sanctioned Under Off-grid Solar Applications of JNNSM." http://www.mnre.gov.in/pdf/jnnsm-offgrid-project. Accessed October 2010.

[380] MNRE. "Test Pilots on Village Energy Security." http://www.mnre.gov.in/adm-approvals/aa-vesp.pdf. Accessed October 2010.

[381] MNRE. *Annual Report 2009-10.* p. 36-37. http://mnre.gov.in/annualreport/2009- 1 0EN/Chapter%204/chapter%204 _1 .htm. Accessed October 2010.

[382] MNRE. *Annual Report 2009-10.* p. 32. http://mnre.gov.in/annualreport/2009- 1 0EN/Chapter%204/chapter%204 1 .htm. Accessed October 2010.

[383] MNRE. "Achievements." http://www.mnre.gov.in/achievements.htm. Accessed October 2010.

[384] Ashkay Urja. "Biomass Gasifier-based Distributed/Off-grid Programme." Vol. 3, Iss. 3, 2009; pp. 15-18. http://www.mnre.gov.in/akshayurja/dec09-e.pdf. Accessed October 2010.

[385] State Nodal Agencies are agencies at the state level that act as implementing agencies for renewable energy programs in their areas. The list of MNRE's state nodal agencies can be found at http://www.mnre.gov.in/list/snas.htm.

[386] Ashkay Urja. "Biomass Gasifier-based Distributed/Off-grid Programme." Vol. 3, Iss. 3, 2009; pp. 15-18. http://www.mnre.gov.in/akshayurja/dec09-e.pdf Accessed October 2010.

[387] ARTI. "Commercialisation of Improved Biomass Fuels and Cooking Devices in India: Scale Up PROJECT." http://www.arti-india.org/index.php?option=com content&view=article&id=44&Itemid=92. Accessed October 2010.

[388] Ashden Awards. "Solar Photovoltaics Enabling Small Businesses to Develop." http://www.ashdenawards.org/files/reports/SELCO2007technicalreport.pdf. Accessed October 2010.

[389] SELCO. "Access to Sustainable Energy Services via Innovative Financing. 7 Case Studies." http://www.selco-india.com/pdfs/selco booklet web.pdf. Accessed October 2010.

[390] SELCO. "Financing Solar Photovoltaic Systems through Rural Finance Institutions." http://enviroscope.iges.or.jp/contents/APEIS/RISPO/inventory/db/pdf/0002.pdf. Accessed October 2010.

[391] WIDA. "Integrated Rural Development of Weaker Sections in India." http://www.indg.in/pdf-files/IRDWSI_casestudy.pdf. Accessed October 2010.

[392] CEA. "The Electricity Act, 2003." http://www.powermin.nic.in/acts notification/electricity act2003/pdf/The%20Electricity%20Act 2003.pdf. Accessed October 2010.

[393] CEA. "All India Electricity Statistics: General Review 2009." New Delhi: CEA, 2009.

[394] CEA. "All India Electricity Statistics: General Review 2009." New Delhi: CEA, 2009.

[395] CEA. "All India Electricity Statistics: General Review 2009." New Delhi: CEA, 2009.

[396] Kalogirou, S.A. "Solar Thermal Collectors and Applications." *Prog. Energy and Combust. Sci.;* Vol. 30, No. 3, 2004; pp. 23 1-295.

[397] MNRE. "Solar Thermal Energy. Thrust Areas for Research and Development." http://www.mnre.gov.in/st-thrust.htm. Accessed October 2010.

[398] United Nations. Division for Sustainable Development. "Agenda 21: Transfer of Environmentally Sound Technology, Cooperation & Capacity-Building." Section IV: Means of Implementation; Chapter 34. United Nations Conference on Environment and Development in Rio de Janeiro: Earth Summit, UN 1992. http://www.un.org/esa/dsd/agenda21/res_agenda21_34.shtml. Accessed August 2010.

[399] Ockwell, D.; Watson, J.; MacKerron, G.; Pal, P.; Yamin, F.; Vasudevan, N.; Mohanty, P. "UK–India Collaboration to Identify the Barriers to the Transfer of Low Carbon Energy Technology." March 2007. http://www.sussex.ac.uk/sussexenergygroup/documents/uk india full pb12473.pdf. Accessed October 2010.

[400] GWEC. "Global Wind Report 2008." http://www.gwec.net/fileadmin/documents/Global%20Wind%202008%20Report.pdf. Accessed October 2010.

[401] GWEC. "Indian Wind Energy Outlook 2009." http://www.indianwindpower.com/pdf/GWEO A4 2008 India LowRes.pdf. Accessed August 2010.

[402] Energy Alternatives India. "India Offshore Wind—Status, Trends, Potential." http://eai.in/blog/2010/04/india-offshore-wind-status-trends-potential.html. Accessed October 2010.

[403] MNRE. "National Solar Mission Document." http://mnre.gov.in/pdf/mission Accessed October 2010.

[404] MNRE. "Jawaharlal Nehru National Solar Mission: Towards Building SOLAR INDIA." http://mnre.gov.in/pdf/mission. Accessed August 2010.

[405] MNRE. *Annual Report 2009-2010.* http://mnre.gov.in/annualreport/2009-10EN/index.htm. Accessed October 2010.

[406] MNRE. "Small Hydro Power Programme." http://www.mnre.gov.in/prog-smallhydro.htm. Accessed October 2010.

[407] Government of India Planning Commission. "Integrated Energy Policy: Report of the Expert Committee." 2006; p. 90. http://planningcommission.nic.in/reports/genrep/rep intengy.pdf. Accessed July 2010.

[408] Kathuria, V. "Technology Transfer for GHG Reduction: A Framework with Application to India." *Tech. Forecasting & Soc. Change*, Vol. 69, 2002; pp. 405–430.

[409] Bridge to India Pvt. Ltd. and Technical University of Munich. "Research Project on Market Barriers for International Renewable Energy Companies in the Indian Market" Research report to be published in 2011.

[410] For detailed information, see http://www.cleanenergyasia.net/. Accessed August 2010.

[411] Ockwell, D.; Watson, J.; MacKerron, G.; Pal, P.; Yamin, F.; Vasudevan, N.; Mohanty, P. "UK–India Collaboration to Identify the Barriers to the Transfer of Low Carbon Energy Technology." March 2007. http://www.sussex.ac.uk/sussexenergygroup/documents/uk india full pb12473.pdf. Accessed October 2010.

[412] U.S. Congress Office of Technology Assessment. "Fueling Development: Energy Technologies for Developing Countries." OTA-E-516. Washington, DC: U.S. Government Printing Office, April 1992.

[413] Ockwell, D.; Watson, J.; MacKerron, G.; Pal, P.; Yamin, F.; Vasudevan, N.; Mohanty, P. "UK–India Collaboration to Identify the Barriers to the Transfer of Low Carbon Energy Technology." March 2007. http://www.sussex.ac.uk/sussexenergygroup/documents/uk india full pb12473.pdf. Accessed October 2010.

[414] Adapted by NREL from Parikh, J.; Kathuria, V. "Technology Transfer for GHG Reduction: A Framework and Case Studies for India." Presented at STAP Workshop; January 19-20, 1997, Amsterdam, The Netherlands.

[415] Copenhagen Economics A/S and The IPR Company ApS. "Are IPR a Barrier to the Transfer of Climate Change Technology?" January 19, 2009; p. 32. http://trade. Accessed August 2010.

[416] Recently the Government of India deregulated energy prices by linking local and international fuel prices. See "Fuel Price Deregulation is Here to Stay." *The Hindu*. 27 June 2010.

[417] India's Cleantech Economy. http://www.indianenergytech.com/2010/03/14/india-increases-national-actionon-climate. Accessed August 2010.

[418] Union Budget. "Key Features of Budget 2010-2011." http://indiabudget.nic.in/ub2010-11/bh/bh1.pdf. Accessed August 2010.

[419] C-Wet. http://www.cwet.tn.nic.in/html/departments_sc.html. Accessed September 2010.

[420] Martinot, E. "Methodological and Technological Issues in Technology Transfer," Chapter 16. *Case Study 22.* http://www.grida.no/climate. Accessed August 2010.

[421] Government of India Planning Commission. "Eleventh Five Year Plan 2007-12: Volume III Agriculture, Rural Development, Industry, Services, and Physical Infrastructure." p. 417. http://planningcommission.nic.in/plans/planrel/fiveyr/11th/11 v3/1 1th vol3.pdf. Accessed August 2010.

[422] MNRE. http://mnre.gov.in/pdf/mission. Accessed September 11, 2010.

[423] Energy Manager Training. http://www.energymanagertraining.com. Accessed September 2010.

[424] BEE. http://www.bee-india.nic.in/. Accessed August 2010.

[425] UNIDO. "Technology Development and Transfer for Climate Change." Working Paper, March 2010, p. 31.

[426] Copenhagen Economics A/S and The IPR Company ApS. "Are IPR a Barrier to the Transfer of Climate Change Technology?" 19 January 2009; p. 5. http://trade 142371 .pdf. Accessed August 2010.

[427] MNRE. "Development of Solar Cities." http://www.mnre.gov.in/pdf/city-guidelines. Accessed September 11, 2010.

[428] UNIDO. "Technology Development and Transfer for Climate Change." Working Paper, March 2010, p. 28.

[429] For detailed information, see http://technology4sme.net/home.aspx. Accessed August 2010.

[430] APCTT. "Renewable Energy Cooperation-Network for the Asia Pacific." http://apctt.net.previewdns.com/solutioncentre/ResAssess.aspx. Accessed October 2010.

[431] UNEP SEFI and Bloomberg New Energy Finance. "Global Trends in Sustainable Energy Investment 2010 – Analysis of Trends and Issues in the Financing of Renewable Energy and Energy Efficiency." 2010; p. 8.

[432] UNEP SEFI and Bloomberg New Energy Finance. "Global Trends in Sustainable Energy Investment 2010 – Analysis of Trends and Issues in the Financing of Renewable Energy and Energy Efficiency." 2010; p. 8.

[433] Ernst & Young. "Renewable Energy Country Attractiveness Indices." p 10. http://www.ey.com/Publication/vwLUAssets/Renewable_energy E/Renewable _Energy _Issue _25.pdf. Accessed September 2010.

[434] UNEP SEFI and Bloomberg New Energy Finance. "Global Trends in Sustainable Energy Investment 2010 – Analysis of Trends and Issues in the Financing of Renewable Energy and Energy Efficiency." 2010; p. 48.

[435] UNEP. "Global Trends in Sustainable Energy Investment 2009. Analysis of Trends and Issues in the Financing of Renewable Energy and Energy Efficiency." 2009; p. 49. http://www.unep.org/pdf/Global trends report 2009.pdf. Accessed October 2010.

[436] UNEP. "Global Trends in Sustainable Energy Investment 2009. Analysis of Trends and Issues in the Financing of Renewable Energy and Energy Efficiency." 2009; p. 49. http://www.unep.org/pdf/Global_trends_report_2009.pdf. Accessed October 2010.

[437] UNEP. "Global Trends in Sustainable Energy Investment 2010 – Analysis of Trends and Issues in the Financing of Renewable Energy and Energy Efficiency." 2010; p. 48. http://bnef.com/DownloadDt3/download/UserFiles/File/WhitePapers/sefi unep global trends 201 0.pdf. Accessed October 2010.

[438] UNEP. "Global Trends in Sustainable Energy Investment 2010 – Analysis of Trends and Issues in the Financing of Renewable Energy and Energy Efficiency." 2010; p. 48. http://bnef.com/DownloadDt3/download/UserFiles/File/WhitePapers/sefi unep global trends 201 0.pdf. Accessed October 2010.

[439] UNEP. "Global Trends in Sustainable Energy Investment 2010 – Analysis of Trends and Issues in the Financing of Renewable Energy and Energy Efficiency." 2010; p. 48. http://bnef.com/DownloadDt3/download/UserFiles/File/WhitePapers/sefi unep global trends 201 0.pdf. Accessed October 2010.

[440] MNRE. "Jawaharlal Nehru National Solar Mission Towards Building SOLAR INDIA." http://mnre.gov.in/pdf/mission. Accessed August 2010.

[441] Commodity Online. http://www.commodityonline.com/news/India-. Accessed August 2010.

[442] Cleantech Group LLC. "Cleantech Venture Capital and Private Equity Investments in India." 2008; p. 15.

[443] IREDA, http://www.ireda.gov.in/. Accessed August 2010.

[444] Cleantech Group LLC. "Cleantech Venture Capital and Private Equity Investments in India." 2008; p. 15.

[445] IREDA. "22nd Annual Report 2008-09." http://www.ireda.gov.in/pdf/IREDA_ANNUAL_REPORT_2008-09.pdf. Accessed August 2010.

[446] The table used extracted information from the official Web sites of PFC http://www.pfc.gov.in/, Rural Electrification Corporation Limited http://www.recindia.nic.in/, and NABARD http://www.nabard.org/. Accessed August 2010.

[447] Cleantech AustralAsia. *Pursuing Clean Energy Business in India* 2008; pp. 23-24. http://www.asiapacificpartnership.org/pdf/REDGTF/RDG-06-10 Report.pdf. Accessed August 2010.

[448] Justice, S.; Hamilton, K. "Private Financing of Renewable Energy—A Guide for Policymakers." http://www.energy guide FINAL-.pdf. Accessed October 2010.

[449] International Finance Corporation. http://www.ifc.org/ifcext/media 57852576EE004C8047. Accessed August 2010.

[450] Ernst & Young. "Renewable Energy in India—Charting Rapid Growth." 2007. https://eyo-iis-pd.ey.com/drivinggrowth/unprotected/downloads/Renewable_Energy_in_India_Charting_Rapid_Growth.pdf. Accessed October 2010.

[451] New Energy India. http://www.newenergyindia.org/Finance%20Page.htm. Accessed August 2010.

[452] ICICI Bank. http://www.icicibank.com/pfsuser/webnews/go-green/Index.html. Accessed August 2010.

[453] Singh, M. "Investors, Banks Willing to Go Green Again." *The Times of India.* http://timesofindia.indiatimes.com/biz/india-business/Investors-banks- 861 547.cms. Accessed August 2010.

[454] More information on this microfinance initiative, http://southasia.oneworld.net/fromthegrassroots/microfinance-for-solar-power. Accessed October 2010,

[455] More information on this microfinance initiative, http://southasia.oneworld.net/fromthegrassroots/microfinance-for-solar-power. Accessed October 2010.

[456] International Finance Corporation. http://www.ifc.org/ifcext/spiwebsite1.nsf/1ca07340e47a35cd85256efb00700cee/72FBCEFEF801C2D7852576 C E0058DB18. Accessed August 2010.

[457] International Finance Corporation. http://www.ifc.org/ifcext/spiwebsite1.nsf/f451ebbe34a9a8ca85256a550073ff10/90a9316abecfa844852577070 0 62fca0?opendocument. Accessed August 2010.

[458] For more information, please visit http://www.german-info.com/press_shownews.php?pid=1725.

[459] For more information on the UNEP Solar Loan Programme, please visit http://www.unep.fr/energy

[460] Tan, V. ADB. Personal Communication. September 29, 2010.

[461] India's participation in CDM is also discussed in Chapter 1.

[462] For a description of Annex I and Non-Annex I countries, see the UNFCCC's parties and observers: http://unfccc.int/parties and observers/items/2704.php.

[463] Institute for Global Environmental Strategies. http://enviroscope.iges.or.jp/modules final.pdf. Accessed August 2010.

[464] Institute for Global Environmental Strategies. http://enviroscope.iges.or.jp/modules. Accessed August 2010. Chart is produced from its original figure.

[465] Institute for Global Environmental Strategies. http://enviroscope.iges.or.jp/modules.

[466] REN21. "Renewables 2010 Global Status Report." Paris: REN21 Secretariat. 2010. http://www.ren21.net/globalstatusreport/REN21 GSR 2010 full.pdf. Accessed October 2010.

[467] Goldman Sachs Global Market Institute. "Alternative Energy: Prospects for Policy, Finance and Technology." June 2009. http://www2.goldmansachs.com/ideas/global-markets-institute/past-research-andconferences/past-research/alt-energy. Accessed October 2010.

[468] REN21. "Renewables 2010 Global Status Report." Paris: REN21 Secretariat. 2010. http://www.ren21.net/globalstatusreport/REN21 GSR 2010 full.pdf. Accessed October 2010.

[469] Europa press release. "Memo on the Renewable Energy and Climate Change Package." http://europa.eu/rapid/pressReleasesAction.do?reference=MEMO/08/33. Accessed September 11, 2010.

[470] REN21. "Renewables 2010 Global Status Report." Paris: REN21 Secretariat. 2010. http://www.ren21.net/globalstatusreport/REN21 _GSR _2010 _full.pdf. Accessed October 2010.

[471] Liming, H. "Financing Rural Renewable Energy: A Comparison between China and India." Institute of South Asian Studies Working Paper. 23 May 2008. http://www.frankhaugwitz.info/doks/general/2008 05 23 China RE Finance Comparison India.pdf. Accessed October 2010.

Indian Renewable Energy Status Report: Background Report for DIREC 2010   121

[472] See Chapters 2 and 4 of this report.

[473] IRENA. http://www.irena.org/ourMission/index.aspx?mnu=mis. Accessed September 11, 2010.

[474] REN21. "Renewables 2010 Global Status Report." Paris: REN21 Secretariat. 2010. http://www.ren21.net/globalstatusreport/REN21 GSR 2010 full.pdf. Accessed October 2010.

[475] Mendonca, M. "Feed-In Tariffs: Accelerating the Deployment of Renewable Energy." 2007.

[476] Hughes, E. "Spain Feed-in Tariff Cuts: New Photovoltaic Solar Power Plants to be Chopped by up to 45%." PV-Tech. http://www.pv-tech.org/news/_a/spain_feed-_in_tariff_cuts_new_phot3 5654654_plants_to_be_chopped_by_up_to_45/. Accessed September 11, 2010.

[477] REN21. "Renewables 2010 Global Status Report" Paris: REN21 Secretariat. 2010. http://www.ren21.net/globalstatusreport/REN21 GSR 2010 full.pdf. Accessed October 2010.

[478] Chan, Y. "South Korea Slashes Import Tariffs on Renewable Energy Equipment." Business Green. http://www.businessgreen.com/business-green/news/2249866/south-korea-slashing-import. Accessed September 11, 2010.

[479] Renewable Power News. "Ways to Encourage Governmental Support for Solar Power." http://www.renewablepowernews.com/archives/1610. Accessed September 11, 2010.

[480] REN21. "Renewables 2010 Global Status Report." Paris: REN21 Secretariat. 2010. http://www.ren21.net/globalstatusreport/REN21 GSR 2010 full.pdf. Accessed October 2010.

[481] Natural Resources Canada. http://www.nrcan.gc.ca/media. Accessed September 11, 2010.

[482] Key Features of Budget 2010-2011. http://indiabudget.nic.in/ub2010-11/bh/bh1.pdf. Accessed September 2010.

[483] REN21. "Renewables 2010 Global Status Report." Paris: REN21 Secretariat. 2010. http://www.ren21.net/globalstatusreport/REN21_GSR_2010_full.pdf. Accessed October 2010.

[484] REN21. "Renewables 2010 Global Status Report." Paris: REN21 Secretariat. 2010. http://www.ren21.net/globalstatusreport/REN21 GSR 2010 full.pdf. Accessed October 2010.

[485] UNEPFI. "The Materiality of Social, Environmental and Corporate Governance Issues to Equity Pricing," Asset Management Working Group, Geneva: UNEPFI; June 2004.

[486] UNEP. "Global Trends in Sustainable Energy Investment 2009." http://www.unep.org/pdf/Global trends report 2009.pdf. Accessed October 2010.

[487] REN21. "Renewables 2010 Global Status Report." Paris: REN21 Secretariat. 2010. http://www.ren21.net/globalstatusreport/REN21_GSR_2010_full.pdf. Accessed October 2010.

[488] Price, S.; Margolis, R. "2008 Solar Technologies Market Report." U.S. Department of Energy. http://www.nrel.gov/docs/fy10osti/46025.pdf. Accessed April 9, 2010

[489] Price, S.; Margolis, R. "2008 Solar Technologies Market Report." U.S. Department of Energy. http://www.nrel.gov/docs/fy10osti/46025.pdf. Accessed April 9, 2010

[490] REN21. "Renewables 2010 Global Status Report." Paris: REN21 Secretariat. 2010. http://www.ren21.net/globalstatusreport/REN21 GSR 2010 full.pdf. Accessed October 2010.

[491] Geothermal Energy Association. "Geothermal Energy: International Market Update." http://www.geo-energy.org/pdf/reports/GEA International Market Report Final May 2010.pdf. Accessed September 12, 2010.

[492] REN21. "Renewables 2010 Global Status Report." Paris: REN21 Secretariat. 2010. http://www.ren21.net/globalstatusreport/REN21 GSR 2010 full.pdf. Accessed October 2010.

[493] DESERTEC. http://www.desertec.org/en/. Accessed September 12, 2010.

[494] MASDAR. "Masdar Turns to Sun's Heat to Cool Buildings." http://www.masdar.ae/en/mediaCenter/newsDesc.aspx?News_ID=150&MenuID=55&CatID=44. Accessed October 14, 2010.

[495] Hawaii Renewable Energy Development Venture. "Five Clean Energy Companies Receive Funding to Demonstrate Innovative Technologies in Hawaii." http://www.hawaiirenewable.com/wpcontent/uploads/2010/09/HREDV-News-release-Sept-201 0.pdf. Accessed October 2010.

[496] Epp, B. "Solar Thermal Based Air-conditioning System Proves Itself in India." http://www.solarthermalworld.org/node/1028. Accessed October 2010.

In: Exploring Renewable and Alternative Energy Use in India   ISBN: 978-1-61209-680-3
Editor: Jonathan R. Mulder   © 2012 Nova Science Publishers, Inc.

Chapter 2

RESOURCE EVALUATION AND SITE SELECTION FOR MICROALGAE PRODUCTION IN INDIA

Anelia Milbrandt and Eric Jarvis

ACKNOWLEDGMENTS

The authors would like to thank the Department of Energy's Office of Biomass Program for providing the financial support for this study. We also thank Dr. Al Darzins and Dr. Philip Pienkos from the National Renewable Energy Laboratory and Dr. H.L. Sharma from India's Ministry of New and Renewable Energy for their review and recommendations. Thanks to the various Indian organizations cited throughout the document for creating and providing the data used in this study.

ABBREVIATIONS AND ACRONYMS

CGWB	Central Groundwater Board
CO_2	carbon dioxide
CPCB	Central Pollution Control Board
FAO	Food and Agriculture Organization
GHG	greenhouse gas
GIS	Geographic Information Systems
MNRE	Ministry of New and Renewable Energy
MPNG	Ministry of Petroleum and Natural Gas
MRD	Ministry of Rural Development
MWR	Ministry of Water Resources
NRSA	National Remote Sensing Agency
PBRs	photobioreactors
R&D	research and development
STP	sewage treatment plant

TAGs	triacylglycerols
TDS	total dissolved solids
WSP	waste stabilization ponds

UNITS OF MEASURE

bbl	barrel
cm	centimeter
ft^3	cubic feet
g	grams
ha	hectare
km	kilometer
km^3	cubic kilometer
kWh	kilowatt-hour
L	liter
m	meter
m^2	square meter
m^3	cubic meter
mg	milligram
Mha	million hectares
MJ	megajoule
mm	millimeter
Mmcf	million cubic feet
Mt	million tonnes
tonne (t)	metric ton

EXECUTIVE SUMMARY

India's growing demand for petroleum-based fuels associated with its growing economy and population presents challenges for the country's energy security given that it imports most of its crude oil from unstable regions in the world. This and other considerations, such as opportunities for rural development and job creation, have led to a search for alternative, domestically produced fuel sources. Biofuels derived from algal oil show considerable promise as a potential major contributor to the displacement of petroleum-based fuels, given its many advantages including high per unit land area productivity compared to terrestrial oilseed crops, utilization of low-quality water sources and marginal lands, and the production of both biofuels and valuable co-products.

The purpose of this study is to provide understanding of the resource potential in India for algae biofuels production and assist policymakers, investors, and industry developers in their future strategic decisions. To achieve this goal, the study integrates relevant resource data from various public and private institutions in India and uses state-of-the-art geographic information systems (GIS) technology to analyze the collected information and visualize the results.

The results of this study indicate that India has very favorable conditions to support algae farming for biofuels production: considerable sunshine, generally warm climate, sources of CO_2 and other nutrients, low-quality water, and marginal lands. Sustainable algae biofuels production implies that this technology would not put additional demand on freshwater supplies and use low-quality water such as brackish/saline groundwater, "co-produced water" associated with oil and natural gas extraction, agricultural drainage waters, and other wastewaters. Although information on the quantity of these water resources in India is not available, the intensity of activities associated with their production suggests that there is a vast potential in the country.

Sustainable algae production also implies that farming facilities would not be located on valuable fertile agricultural lands but on marginal lands (classified as wastelands in India). These lands include degraded cropland and pasture/grazing land, degraded forest, industrial/mining wastelands, and sandy/rocky/bare areas. It is estimated that the extent of these lands in the country is about 55.27 Mha, or approximately 18% of the total land area. If India dedicates only 10% (5.5 Mha) of its wasteland to algae production, it could yield between 22–55 Mt of algal oil, which would displace 45%–100% of current diesel consumption. The production of this amount of algae would consume about 169–423 Mt of CO_2, which would offset 26%–67% of the current emissions.

The study illustrates options for siting two algae production facility types: *co-located facilities*, which produce algal biomass as a consequence of another process, and *dedicated facilities*, with the main purpose of algae production. Co-located facilities are those operating in conjunction with wastewater treatment where algae are produced as a byproduct of the wastewater treatment process. Using the wastewater effluent as pond medium provides a cost-effective solution to water, land, and nutrients considerations because the wastewater treatment function would cover nearly all costs. Domestic and industrial wastewater treatment facilities were considered in the analysis as potential sites for co-locating algae farms. This opportunity exists in most states of India.

The site-suitability analysis for dedicated algae farms was centered on the stationary industrial CO_2 sources given key advantages of co-locating algae production with these facilities: supplying carbon for enhanced algal growth with low/no transportation costs and a means for capturing CO_2 before it is released to the atmosphere, thus providing potential carbon credits for utilities. The study considered stationary CO_2 sources in areas where these facilities coincide with other inputs necessary for algae growth or conditions that meet the engineering, economic, environmental, and social requirements for this technology. The results of this analysis indicate that suitable locations for dedicated algae farms are in the western and southern parts of the country and along coastal areas.

India is a large country with diverse landscape and resources. Therefore, future work could focus on a state or even smaller geographic area to provide a more detailed examination of the resource potential for algae production and pinpoint the most suitable locations. The authors believe that the information provided in this study will serve as a base for further analysis of the algae biofuels potential in India and assist policymakers and industry developers in their strategic decisions.

INTRODUCTION

India's growing demand for petroleum-based fuels associated with its growing economy and population presents challenges for the country's energy security given that it imports most of its crude oil from unstable regions in the world. This and other considerations, such as opportunities for rural development and job creation, have led to a search for alternative, domestically produced fuel sources. Biomass-derived fuels, namely biodiesel and ethanol, are considered by the Indian government as one option to substitute petroleum-based products—diesel and gasoline, respectively. To promote the production and use of these fuels, the government announced a Biofuels Policy in December 2009 that calls for the blending of at least 20% biofuels with diesel and gasoline by 2017 (MNRE 2009).

India is essentially a diesel-driven economy. Diesel consumption in 2008–2009 was 52 million tonnes (Mt)—about 40% of the total petroleum products consumed—against gasoline consumption of 11 Mt (MPNG 2009), and it grows by about 6%–8% annually. Diesel is widely used in all sectors—transportation, agriculture, power generation, and industry—but it is mainly consumed in road and rail transport (more than 50% of total diesel use). Biomass-based diesel substitutes include: *biodiesel*, typically produced by chemically reacting lipids (vegetable and waste oils such as animal fat and used cooking oil) with an alcohol; *green diesel*, produced either from hydroprocessing lipids or indirect liquefaction of any biomass feedstock[1] ; and *unmodified vegetable and waste oils*, which may be used directly, without conversion, in diesel engines. Feedstock for these fuels considered in India (except green diesel produced from indirect liquefaction, which could use any source of biomass) includes non-edible vegetable oils such as those from Jatropha (*Jatropha curcas*) and Karanj (*Pongamia pinnata*) and waste oils. In addition, biofuels derived from algal oil show considerable promise as potential major contributors to the displacement of petroleum-based fuels, given the many advantages including high per unit land area productivity compared to terrestrial oilseed crops, utilization of low- quality water sources (brackish, saline, and wastewater) and marginal lands, and the production of both biofuels and valuable co-products.

The purpose of this study is to provide understanding of the resource potential in India for algae biofuels production and assist policymakers, investors, and industry developers in their future strategic decisions. To achieve this goal, the following two project objectives were defined:

1. Examine the resources available for algae production in India.
2. Identify areas suitable for algae production in the country.

To accomplish these objectives, the study integrates relevant resource data from various public and private institutions in India and uses state-of-the-art geographic information systems (GIS) technology to analyze the collected information and visualize the results.

BIOFUELS FROM MICROALGAE: OPPORTUNITIES AND CHALLENGES

Microalgae for Lipid Production

The term "algae" refers to a very large group of photosynthetic, aquatic organisms that lack the true roots, stems, and leaves of higher plants. Eukaryotic algae are generally divided into the multicellular "macroalgae" (such as seaweeds) and the unicellular "microalgae." Algae represent an enormous amount of biodiversity, with over 40,000 species characterized so far. These organisms are remarkably adaptable and occupy virtually every environment on the planet. They can be found in waters of widely varying temperature, pH, and salinity (from freshwater to hyper-saline). For some time it has been known that many species of microalgae have the ability to accumulate lipids in the form of triacylglycerols (TAGs). This oil can reach 20%–60% of the dry cell weight and is typically synthesized in response to conditions of stress, such as lack of nitrogen. This response allows cells to store carbon and energy during times when cell division is limited. Of the 12 or more major divisions of algae, the best studied for this oil accumulation phenomenon fall into the Chlorophyceae (green algae) and Bacillariophyceae (diatoms).

The worldwide interest in exploiting oleaginous algae for their ability to make oil has reached a frenzied pace in recent years. However, the basic concept of growing algae for fuels has been around for more than 50 years (Meier 1955). Under the U.S. Department of Energy's Aquatic Species Program, research in this area accelerated the field with over $25 million and 18 years spent isolating algal strains from the wild, characterizing their physiology and biochemistry, developing genetic systems, exploring downstream processes, demonstrating mass culture systems, and analyzing the economics and resource requirements. A comprehensive overview of the program was published (Sheehan et al. 1998). The conclusion from these studies was that the technology is extremely promising, but major advances will be required to make it a cost- effective source of fuel, particularly given the fact that petroleum was trading for less than $20/barrel in 1995 when the program was winding down.

Benefits of Microalgal Oil Production

Why has the potential of producing fuels from microalgae been receiving so much attention in recent years? The answer lies both in externalities, such as recent escalations in the price of oil, and the inherent attractiveness of the technology. Microalgae can be grown with minimal inputs: land, sunlight, water, some macro- and micro-nutrients, and carbon dioxide (CO_2). The land need not be fertile, productive land; the ability to grow algae in wasteland regions means that the technology does not compete directly with food cropping. Similarly, low quality water is also applicable. Species of microalgae can be found that thrive in a wide range of salinities, thus brackish and saline water resources that are unsuitable for traditional crops will work well for microalgae cultivation. The requirements for macro- and micro-nutrients are similar to those of higher plants—nitrogen, phosphorous, and iron are key. Because these elements do not end up in the fuel product, there is the potential to recycle most of these nutrients in the process. Another avenue is the potential for domestic, industrial, and agricultural wastewater streams to be used as both water and nutrient sources for algal cultivation, which has the added benefit of remediating these waste streams. Finally, it is important to provide CO_2 at elevated concentrations for productive growth. Algae can be grown using industrial CO_2 streams, such as flue gas from electricity generation stations (typically 10%–1 5% CO_2). Thus, there is the potential to provide the algae's key nutrient (CO_2) while capturing and recycling the primary gas responsible for global warming.

Given the right resources, microalgal oil productivities can be quite high. This is a result of their rapid growth rates (often > 1 doubling per day) and their potential for high oil content. Land and water resource limitations associated with using traditional oil crops, such as soybean or oil palm, prevent them from ever providing a large fraction of the world's demand for diesel and aviation fuels. Microalgae, on the other hand, can be conservatively expected to produce two times the oil yield per hectare of oil palm or 25 times the oil yield per hectare (ha) of soybean. This means that the land impact becomes reasonable for making a very significant contribution to the supply of hydrocarbon fuels. The oil (TAGs) from microalgae typically contains fatty acids that are very similar to those of conventional oil crops and can be converted readily to biodiesel or other fuels.

Compared to conventional petroleum-based fuels, the energetics and greenhouse gas (GHG) emissions associated with the production of microalgal fuels have the potential to be very attractive. The ultimate source of energy in these fuels is the sun, captured by the photosynthetic machinery of the algal cells. The carbon in the fuels is captured from waste streams, such as power plant flue gas; thus, a second use of this carbon is achieved before release to the atmosphere. Early life cycle analysis (LCA) studies, however, indicate that the overall energy and GHG balance is very sensitive to the design of the process (e.g., Huesemann and Benemann 2008; Lardon et al. 2009). Since an active industry producing microalgal biofuels does not yet exist, it is difficult to know the appropriate inputs and assumptions to use in such LCA modeling. The source of key nutrients and the need for drying algal biomass before oil extraction are particularly important determinants of the LCA results. Utilization of wastewater can improve the LCA, as can methods of extraction that do not require drying the biomass. These and other factors must be carefully considered before any large-scale implementation of the technology.

Algal Cultivation Systems

Microalgae can be cultivated in any system that provides the right environment for growth, including sufficient light, gas exchange (CO_2 delivery and O_2 removal), temperature control, and mixing. The ultimate goal is to optimize all of these parameters so as to approach the theoretical efficiency of energy capture through photosynthesis, which is difficult in practice. Balanced against providing the ideal growth environment is the system cost, which must be kept to a minimum. Proposed growth systems vary from simple open ponds to complex photobioreactors (PBRs). Growth systems are currently an area of active research worldwide.

Open ponds are the most cost-effective algal growth systems. Unstirred ponds tend to have very low productivity rates. Raceway designs, however, which use paddlewheels to provide laminar flow of the culture and keep the algae suspended, can be very productive.

These shallow, artificial ponds are typically 15–30 cm deep and can be lined with clay or plastic to prevent percolation through the bottom. CO_2 is generally sparged into the culture in a deeper section known as the sump. These systems scale up very well and have the advantage of less severe temperature swings due to good thermal contact with the ground and evaporative cooling from the surface.

There are many designs for PBR growth systems, the most common of which fall into the categories of flat panel, tubular, and vertical column systems (for reviews, see Richmond 2004; Chisti 2007; Eriksen 2008; Ugwu et al. 2007). By optimizing the surface-to-volume ratios, PBRs can offer higher volumetric productivities than ponds, but not necessarily higher areal productivities. They have the potential for better culture stability due to decreased risk of contamination, although at large scale it is impossible to keep such systems axenic. Water consumption can be reduced relative to open ponds due to decreased evaporation, unless evaporative cooling needs to be employed for temperature control. PBRs suffer from the complexities of gas exchange (CO_2 introduction and O_2 removal), mixing, temperature control, and prevention of fouling (biofilm formation). But even more critical is the capital costs of such systems because of their complexity and the expensive materials used for construction (e.g., steel, glass, and plastic). Some authors have predicted that the increased performance of PBRs will never compensate for the added economic and life cycle costs of such systems (Huesemann and Benemann 2008; Sheehan et al. 1998).

Hybrid systems have also been proposed in which a combination of open ponds and PBRs are used to optimize growth and lipid induction. Heterotrophic growth, in which algae are fed biomass-derived sugars, is also an option; however, this approach relies on agricultural feedstocks that have other uses. Artificial light systems have been proposed, allowing better control of temperature and sterility, but such concepts are clearly not feasible due to energy and cost considerations. Finally, offshore cultivation is an option to consider, although reliable infrastructure for such systems has not yet been devised. For the purposes of this paper we will consider only land-based, phototrophic systems.

Downstream Processing to Fuels

Productive growth of oil-rich algae is only half of the equation. Equally important is the harvesting of the algal biomass, extraction and purification of the oil (TAGs), and conversion to usable transportation fuel. All three steps are important to the overall economics and the subject of ongoing research.

Microalgal harvesting, or "de-watering," has the potential to be a very cost- and energy-intensive process. Even a "dense" algal culture may only contain 1 g of dry cell weight per liter of culture (i.e., 0.1% solids). A 1 hectare, 20 cm deep open pond would contain 2,000 m^3 (2×10^6 liters) of culture, and it may be necessary to harvest half of the pond volume (1 million liters) every day. The harvesting system must therefore be rapid and efficient. Some of the options being considered include spontaneous settling of the cells, autoflocculation, bioflocculation, chemical flocculation, centrifugation, filtration, and dissolved air flotation. In reality, harvesting will likely include a combination of such techniques, such as using flocculation to achieve 1 %–2% solids, followed by centrifugation or tangential flow filtration to achieve 20% + solids. Many of these methods are somewhat strain-dependent, and more research is needed in this area.

The TAGs in microalgal cells are contained in oil droplets within the cells, and the cell wall can pose a significant barrier to removal of the oil. Extraction of algal oil probably involves three steps: 1) disruption of the cell wall, 2) separation of the oil from the remaining biomass, and 3) purification or upgrading of the oil to remove impurities. Many approaches to extraction are being studied. Solvent extraction, using hexane, for example, can be used successfully to extract oil from intact cells of many species, and the algal biomass need not be fully dried. Drying the algae is extremely energy intensive and must be avoided if possible; ideally a technique will be able to deal with an aqueous slurry or paste (e.g., 20% solids). Solvent extraction, however, can be costly and pose serious environmental concerns. Other approaches being studied involve more benign ("green") solvents, supercritical CO_2, enzymatic extraction, and ultrasonic cell disruption, among others. Again, methods can be very strain-specific, and much more research is needed to find inexpensive, safe, and scaleable methods for oil extraction.

Finally, the extracted TAGs need to be converted into a fuel that is compatible with the existing transportation infrastructure. This means converting the oil into diesel or aviation fuel substitutes that meet all of the relevant specifications for fuel quality. Two main pathways are being considered. The first is the conversion of the TAGs into alcohol esters (i.e., biodiesel) using conventional transesterification technology. The second is to use catalytic hydroprocessing methods to generate a renewable "green" diesel product which does not contain oxygen. Both approaches have their merits and are based on well-understood processes. The hydroprocessing approach, which essentially uses conventional oil refinery operations, has the benefits of more process flexibility and the ability to address aviation fuel markets as well as diesel. The chosen process must also be able to deal with variability in the oil feedstock, as the fatty acid composition of oils can vary based on algal species and growth conditions.

Economics

The technical feasibility of algal biofuels has been demonstrated. The challenge comes in making the system cost-competitive with other fuel sources. While government incentives and monetization of externalities (e.g., environmental benefits) can help to give this nascent technology a boost, becoming directly competitive with the cost of petroleum-derived fuels is the best way to ensure growth of an algal biofuels industry. Unfortunately, current estimates of the cost of algal-derived biofuels vary over two orders of magnitude, depending upon how optimistic the assumptions are. Until large-scale facilities are operational and we have a better understanding of the performance and production costs, it is difficult to make accurate assumptions in cost models. Probably the most important determinant of the economics is the biological productivity, that is, using a productive strain of microalgae and optimizing the growth system (Sheehan et al. 1998). However, other factors such as the cost of CO_2 and water and the harvesting and extraction systems also play important roles. Other inputs, such as the cost of land, are somewhat less significant.

One of the major areas of uncertainty in the cost modeling is the value to be gained from co- products. Even if cells are 50% oil, it means that the residual biomass comprises another 50% of the mass. This is at worst a disposal issue and at best a revenue opportunity of even more value than the oil itself. Most microalgal cells are quite high in protein, which has potential applicability to food and feed markets. Residual carbohydrates can also have feed value or could be fermented to ethanol as a gasoline substitute. Anaerobic digestion of the entire residual biomass could produce biogas for heating and power generation. Higher value products can also be found in many species of algae, including pigments (e.g., carotenoids and astaxanthin) and specialty lipids (e.g., omega-3 fatty acids). Co-products are critical to the overall economic viability of algal biofuels, and care must be taken to match the size of the co-product market to the projected size of algal fuel production facilities.

Challenges to Overcome

Despite the promise of algal biofuels, sustained yields obtained so far for algal mass culture efforts have fallen significantly short of the levels required for cost-effective fuel production (Hu et al. 2008). Further strain improvement and culture optimization may allow the necessary yields to be achieved. However, there are challenges throughout the process chain that need to be addressed.

Large-scale cultivation of microalgae is still in its infancy. First, robust, highly productive strains must be identified that exhibit consistently both high growth rates and high oil content. The stability of cultures is critical, as culture "crashes" will take a severe toll on overall productivity. Stability will be affected by invasion of weed algal species, pathogens (e.g., viruses and fungi), and predators/grazers that feed on microalgae. CO_2 supply, gas exchange, and temperature control could be problematic in many situations. Nutrient sourcing and recycle can be critical to the economics. The overall cost of cultivation systems, especially the capital infrastructure costs, must also be reduced.

The downstream processes also pose significant challenges. As discussed above, the harvesting and extraction of algal cells are critical cost components for which economically viable solutions may not yet exist. The characteristics, stability, and consistency of the final fuel product need to be addressed. Finally, the management of water throughout the process is of paramount importance, particularly in arid regions. This includes sourcing water for cultivation systems, makeup water to replace losses due to evaporation, and treatment of incoming and outgoing water streams (e.g., removing nutrient residues, organics, heavy metals, salt, and live organisms). Of course, the regulatory requirements, environmental impacts, and life cycle benefits of the entire process must also be carefully considered.

It is clear that one of the keys to the success of this industry is to align the necessary resources at the required costs. The benefits of the technology will not be realized unless the process can be made economical and the overall benefits shown to be beyond dispute. Although demonstration and pilot studies are now beginning, the only way this technology will have an impact on providing transportation fuels in a sustainable and cost-effective manner is for the required resources to be available to cultivate microalgae on a truly massive scale. For that reason, siting and resource issues are at the fore. India's capacity to provide those resources for a large-scale and sustainable algae-to-biofuels industry is the focus of this paper.

RESOURCE EVALUATION

The decision for locating an algae production facility in a given area, as with many other technologies, starts with evaluating the resource potential. Favorable climate conditions and availability of water, CO_2, and other nutrients (primarily nitrogen and phosphorous) must be aligned with suitable land characteristicstopography, soil, and use—to determine the most promising areas for algae production. Because these resources vary considerably from one geographic location to another, optimal siting of algae farming systems requires knowledge of the specific resource characteristics—availability, magnitude, and variability—at any given location. An overview of the critical resources for algae production systems in Indiaclimate, water, CO_2, other nutrients, and land is presented below.

Climate

The impact of climate on algae production is comparable to the impact of climate on terrestrial plants. Like higher plants, algae need abundant sunlight and have differing tolerances to temperature. Closed photobioreactors can be less sensitive to climate variability than open pond systems due to the somewhat more controlled environment they provide. Solar radiation and the number of daily sunshine hours directly affect algae productivity, precipitation and evaporation affect water supply, and severe weather (hail, dust storms, and floods) impacts water quality.

Equally important to both open and closed algae cultivation systems is the availability of abundant sunlight. Figure 1 illustrates the annual average global horizontal solar radiation in India. Global solar radiation is the sum of the direct, diffuse, and ground-reflected radiation arriving at the earth's surface, also called total solar radiation. *Direct* beam radiation comes in a direct line from the sun; *diffuse* radiation is scattered out of the direct beam by molecules, aerosols, and clouds; *ground-reflected* is the solar radiation reflected back into the atmosphere after striking the earth. Solar radiation of 4.0 kWh/m^2/day (approximately $14MJ/m^2$) is considered adequate for algae production. Therefore, almost the entire country (except parts of Arunachal Pradesh) is suitable for algae production from the standpoint of receiving sufficient solar radiation on an annual basis. There are, however, monthly variations as illustrated in Figure 23 (Appendix). Lower solar radiation values are experienced during the monsoon season in the central, northeastern, and western coastal states and during the post-monsoon season/early winter in the northern states. Algae productivity in those locations would be low during these months.

Another important climate element affecting algae growth is the number of sunshine hours during the day. In most parts of India, clear sunny weather is experienced 250 to 300 days per year with the sunshine hours ranging between 2,300 and 3,200 per year depending upon the location (Muneer, Asif, and Munawwar 2004). Figure 2 illustrates the annual average daily sunshine hours in India. The values vary from less than five hours for locations in the northeastern states to more than nine hours for places in the western and central states.

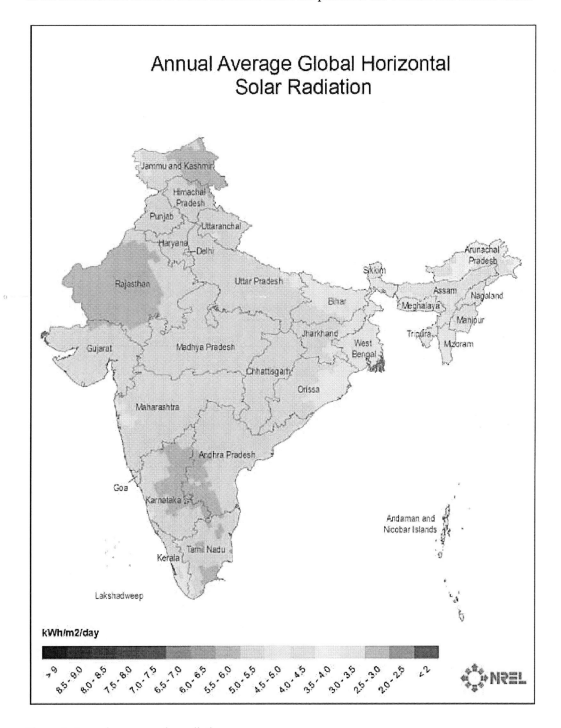

Figure 1. Annual average solar radiation.

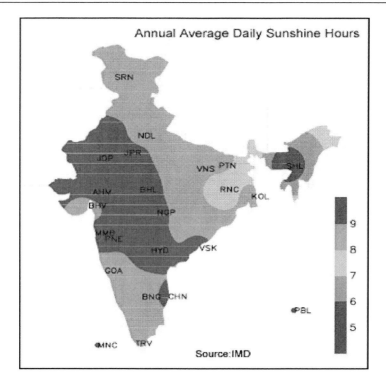

Figure 2. Annual average daily sunshine hours.

The India Meteorological Department (IMD 2009) designates four official seasons: winter (January through March), summer or pre-monsoon season (April through mid-June), monsoon season (mid-June through September), and post-monsoon season (October through December). On average, the lowest number of daily sunshine hours is experienced during the monsoon months—for some locations less than four hours. During the other seasons, the sunshine hours are high (more than eight hours) for most of the country. The monthly variations are shown in Figure 24 (Appendix). Days with less than six hours of sunshine are considered inadequate for algae growth; therefore, productivity would be low during the rainy season in some areas of the country. Although algae productivity would be low at these locations during the monsoon months, the remaining 8 to 9 months of the calendar year provide a sufficiently long growing period for algae.

The various species of microalgae grow under a wide range of temperatures; thus, temperature per se is not a critical climate element. However, most algae species are sensitive to freezing; therefore, areas reaching 0°C (32°F) and below in the winter months would be challenging locations for algae farming. Thermal mass of the cultivation system and high salinity growth medium might protect microalgae during temperature excursions below 0°C, but productivity would be negligible under such conditions. Areas prone to freezing weather include the mountain areas in Northern India: the states of Jammu and Kashmir, Himachal Pradesh, Uttaranchal, Sikkim, and Arunachal Pradesh (Figure 3). Portions of these states, at the sub-Himalayan and sea-level elevation, generally experience subtropical climate with mild winters, but temperatures even in these areas may fall to 0°C during the coldest months. Similarly, temperatures in some areas of western Rajasthan can drop below freezing due to waves of cold air from central Asia during winters. If algae farming is considered in these areas, the facilities may need to be equipped with a heating mechanism to keep the ponds at an optimal temperature for algae growth. This, however, would add substantially to the farms' operating costs and defeat the purpose of algal fuels production, which is to reduce GHG emissions, if fossil fuel sources are used. Alternative sources of energy could be considered to improve the environmental impact of these facilities, such as geothermal resources (available in areas with cold temperatures as shown in Figure 25; Appendix), but this approach may not be cost competitive. Geothermal resources are being researched in the United States as a heat source for algae farms, particularly in Nevada (a state with wide local temperature variations), but the cost of this method is yet to be evaluated. Another option is to consider waste heat from power plants and other CO_2 emitting industrial facilities.

As mentioned earlier, precipitation and evaporation do not affect algae productivity directly. However, they are important climate elements to consider in siting algae farms as they relate to water supply and loss. Precipitation affects water availability (surface and groundwater) at a given location. The IMD estimates the annual average rainfall in India during 2000–2005 at 1,117 mm (Hindustan Times 2005). More than 75% of the rainfall is received during the four monsoon months (Harihara Ayyar 1972). Of the remaining volume, a large fraction is received during the winter months. Areas with sufficient precipitation (more than 1,000 mm/year) include northeastern states, western Ghats, central states, and the eastern part of Ganges Valley (Figure 4). A recent study by the Indian Tropical Meteorology Institute in Pune looked at data from 1871 to 2006 and concluded that the rainfall patterns in the country are changing: rainfall is decreasing along the Ganges Valley and in the northeastern states (Figure 5; Singh et al. 2005). Rainfall is increasing in West Bengal, Jharkhand, Assam, parts of Gujarat (historically dry), and most coastal regions of the country. The observed rainfall trends will affect water abundance/deficit in those regions; therefore, it will have an impact on the water available for algae production.

Evaporation contributes to water loss. It is a very important factor to consider when choosing locations for open pond farming because it can significantly increase operating costs. This is less of an issue for closed PBRs, although evaporation is necessary to help regulate the temperature of both open and closed systems. The western, central (Maharashtra), and southern (Tamil Nadu) states in India have the highest evaporation rates in the country with more than 275 cm annually (Figure 6). Evaporation rates follow closely climatic seasons: they are low during the winter (2–6 mm/day), monsoon (4–10 mm/day), and post-monsoon (4–8 mm/day) months and reach their peak in the summer months of April and May (5–16 mm/day) (NIH 2009). Saline water surfaces evaporate less than freshwater surfaces. Therefore, algae ponds using brackish or saline water should be considered in areas where the evaporation rates are high.

Severe weather can have a major impact on aquaculture. In India, natural disasters that could be devastating to algae farming include cyclones along the coastal areas, hail and dust storms in northern states, widespread floods along the rivers, and flash floods during the monsoon season (Figure 26 and Figure 27; Appendix). These weather events could contaminate the open pond environment or damage covered systems; therefore, the potential for such natural disasters should be taken into account when looking at prospective sites for algae production in India.

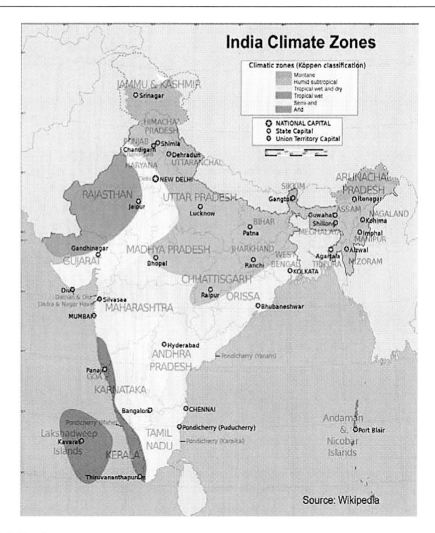

Figure 3. India climate zones.

Water Resources

The renewable water resources in India are estimated at 1,869 km^3 annually (MWR 2009). The country ranks ninth in the world; in comparison, Brazil has more than 8,000 km^3, Russia has 4,500 km^3, and Canada and the United States have more than 3,000 km^3 each (CIA 2009). Due to topographic, hydrological, technological, and other constraints, however, the amount of water that can actually be utilized in India is less—about 1,123 km^3. Of this amount, surface water represents 61% or 690 km^3, and groundwater is about 433 km^3; although, considering natural discharge during non-monsoon season, groundwater availability is actually closer to 399 km^3/year.

India has more than 20 major river systems providing irrigation, drinking water, transportation, electricity, and the livelihoods for many people all over the country as illustrated in Figure 7. The rivers of India are broadly classified into four groups: Himalayan rivers, Deccan rivers, coastal rivers, and rivers of the inland drainage basin. Some of these rivers are perennial and others seasonal (Figure 8). The Himalayan rivers—Ganges, Brahmaputra, and Indus—are formed by melting snow and glaciers, and therefore, continuously flow throughout the year. Together, these river basins provide about 50% of the surface water resources in the country (Kumar et al. 2005). Most of the Deccan rivers (e.g., Godavari, Krishna, Cauvery, Mahanadi, Narmada, and Tapti) and coastal rivers are seasonal; their flow depends mainly on rainfall. Therefore, adequate rain during the monsoon season is critical in these parts of the country for a lack of it leads to water shortage. The streams of inland drainage basins are small rivers in sandy areas of Rajasthan with no outlet to the sea, except Luni, and most of them are of an ephemeral character.

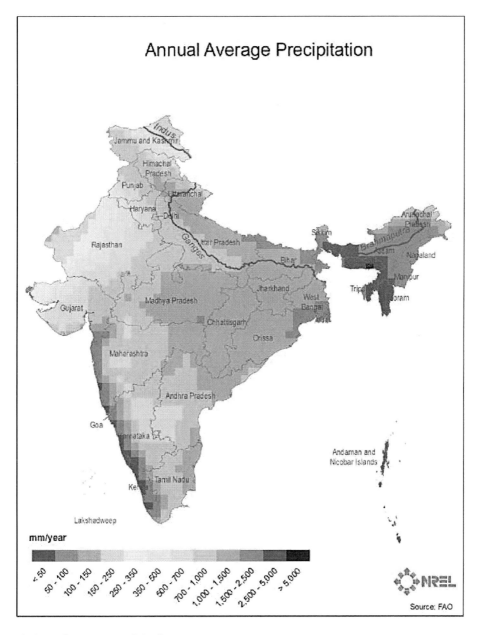

Figure 4. Annual average precipitation.

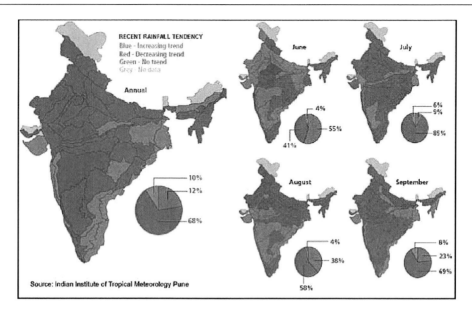

Figure 5. Rainfall trends, 1813–2006.

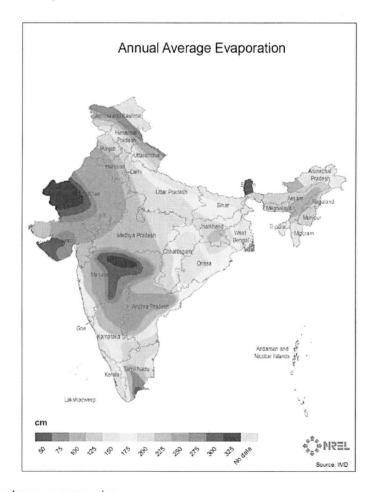

Figure 6. Annual average evaporation.

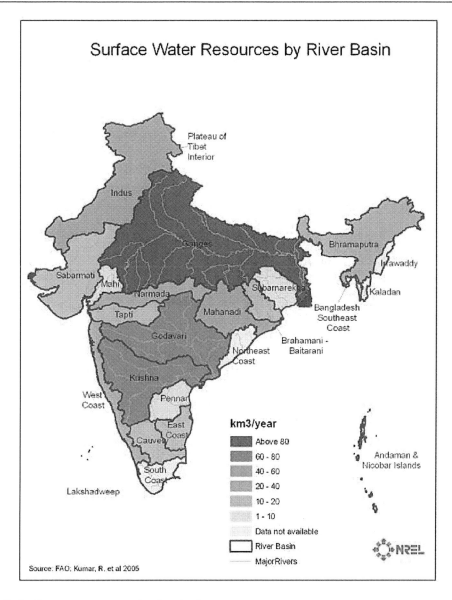

Figure 7. Surface water resources by river basin.

In addition to surface water, groundwater is also an important resource for irrigation, domestic, and industrial uses in India. It is a strategic source of freshwater during dry periods. It accounts for about 80% of domestic water use and more than 45% of the irrigation in the country (Kumar et al. 2005). Figure 9 illustrates the groundwater resources in the country by district. Districts along the main rivers in the north and in the southern states of Andhra Pradesh and Maharashtra have the highest volume of groundwater resources.

Groundwater in India is recharged by rainfall (67%) and other sources (canal seepage, return flow from irrigation, seepage from water bodies, and artificial recharge due to water conservation structures). In some areas of the country, groundwater levels have been declining over the past years. This is due to a number of factors—decreased rainfall, deforestation, and soil degradation—but is particularly due to overexploitation resulting from population growth, economic development, and water-intensive farming. The Central Ground Water Board (CGWB) of the Ministry of Water Resources (MWR) evaluated the groundwater resources at the finest administrative unit (blocks, talukas, and mandals) and categorized them into several groups as illustrated in Figure 10 (CGWB 2006a). Groundwater in about 20% of all administrative units— predominantly in the northwestern, western, and southern parts of India—is categorized as "overexploited" and "critical." Approximately 10% of all units have groundwater that is categorized as "semi-critical" and about to reach critical condition if proper management of these resources doesn't take place. Scientists from the U.S. National Aerospace and Space Administration (NASA) studied satellite images of groundwater storage from 2002 to 2008 in north India and found that the groundwater levels in north India are declining by 33 cm per year and that about 110 km^3 of groundwater have been lost over the six-year study period—double the capacity of India's largest surface water reservoir, the Upper Wainganga (NASA 2009).

Figure 8. River flow patterns.

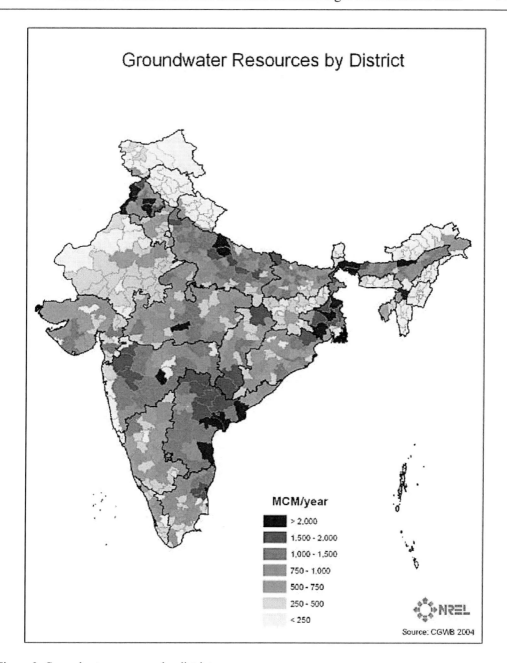

Figure 9. Groundwater resources by district.

Overexploitation of groundwater resources, more so than climate variations and land use change, causes a lowering of the water table, which makes the groundwater more difficult and expensive to obtain. In many parts of India, groundwater is increasingly being pumped from lower and lower levels. In some areas, well drillers are using modified oil-drilling technology to reach water, going as deep as 1,000 m (Brown 2007). Groundwater level fluctuations (the rise and fall of the water table) over a decade are shown in Figure 11. The areas where the water table is falling overlap with highly populated places and locations with groundwater overexploitation.

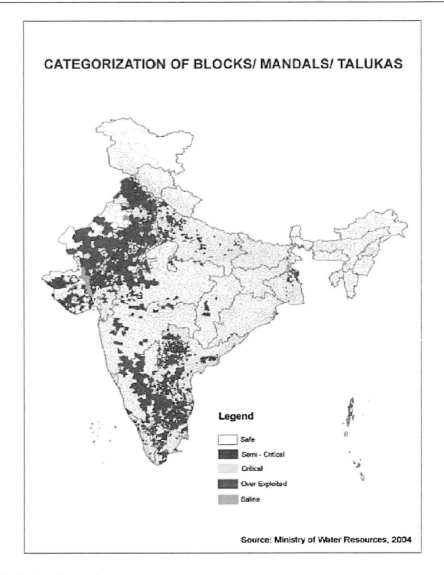

Figure 10. Status of groundwater resources.

Because depth to water table greatly affects the cost of extraction, this will negatively impact the economics of growing algae in those areas. Naturally, locations with groundwater closer to the surface would provide a cost effective way to produce algae. It is critical that the emerging algae industry be wary of regions where added consumption will exacerbate already serious water resource issues.

One of the benefits of growing algae is that it can utilize low-quality water with few competing uses, such as brackish/saline groundwater, "co-produced water" from oil and natural gas wells, and wastewater discharged from domestic, industrial, and agricultural activities. Therefore, algae technology need not put additional demand on freshwater supplies. Using poor quality water for algae production can serve two purposes: one is to dispose of water that would otherwise be costly (economically and environmentally) to dispose, and the other is to utilize poor quality water to create products that have economic value—algal biofuels and biomass for animal feed or fertilizer.

Resource Evaluation and Site Selection for Microalgae Production in India 143

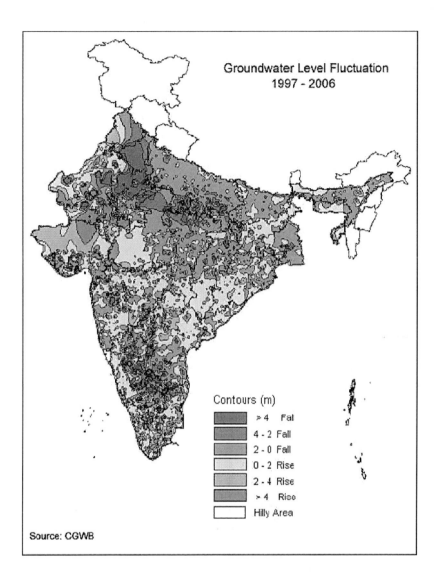

Figure 11. Groundwater level fluctuation.

Brackish water is defined as water containing total dissolved solid (TDS) concentrations between 1,000 and 10,000 milligrams per liter (mg/L). This definition includes slightly-saline (1,000 to 3,000 mg/L TDS) and moderately-saline (3,000 to 10,000 mg/L TDS) water (USGS 2009). As a reference, freshwater contains less than 1,000 mg/L, highly-saline water has about 10,000 to 35,000 mg/L TDS (seawater has a salinity of roughly 35,000 mg/L TDS), and brine has more than 35,000 mg/L TDS. Groundwater with TDS concentrations greater than 3,000 mg/L is unusable for irrigation without dilution or desalination and is not safe for most poultry and livestock watering (Warner 2001).

High salt levels occur naturally in some parts of India such as the coastal areas, but in many cases this has been exacerbated where human activities accelerate the mobilization and accumulation of salt. Inland groundwater salinity is present in arid and semi-arid regions of the country. Figure 12 illustrates the geographic distribution of brackish groundwater in India.

There are several places in Rajasthan and southern Haryana where TDS in groundwater exceeds 6,700 mg/L making the water unsuitable for agricultural activities without treatment (CGWB 2006b). Desalination of saline or brackish water for use in irrigation and human consumption, in addition to being a costly option, presents some environmental challenges associated with its energy- intensity, GHG emissions, and disposal of the salty brine by-product. Many species of algae have a high tolerance to salinity; therefore, in addition to coastal areas, the algae technology can target parts of the country where the presence of saline groundwater resources prevents their direct use in other applications.

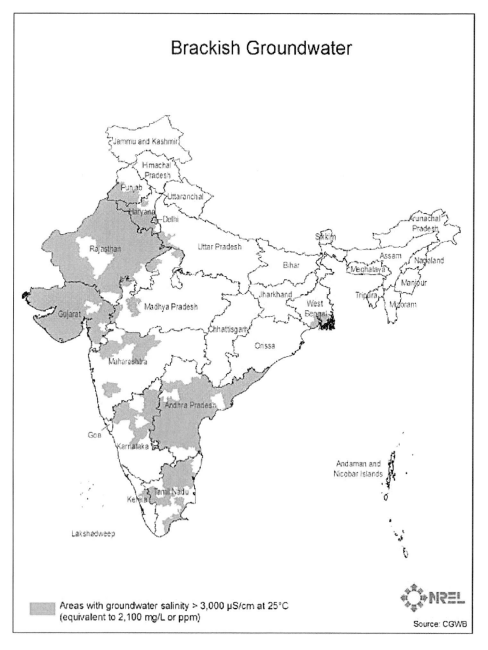

Figure 12. Brackish groundwater in India.

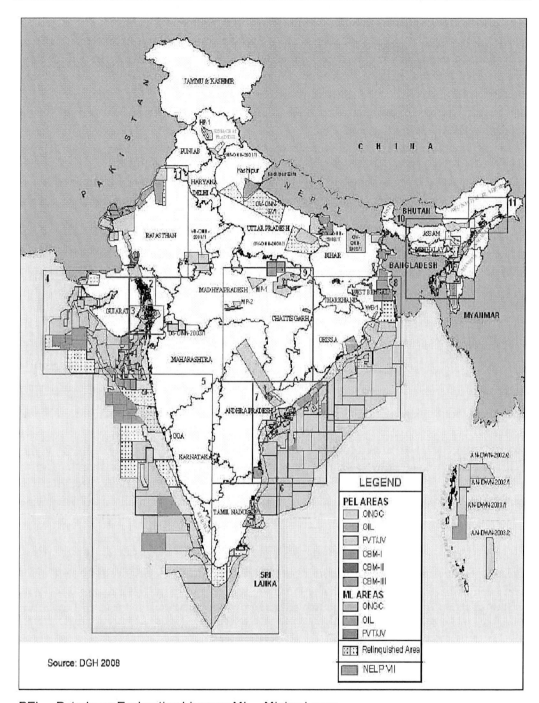

PEL – Petroleum Exploration License; ML – Mining Lease

Figure 13. Oil and gas fields in India.

If brackish/saline water is used for algae production, an important consideration is where and how algal growth media is disposed of after cultivation. Although the water can be used for many generations of algal production, evaporation (particularly in open ponds) will concentrate higher levels of salts in the water and ultimately to beyond the range suitable for algal growth. The typical method for dealing with this problem is to dispose of a certain amount of the spent media during harvesting as "blowdown" water. This water is then replaced with new water from the source (fresh or saline). The blowdown water will contain more elevated levels of salts and potential contaminants from flue gas, algal byproducts, and probably some live algal cells. Returning this water to the source (e.g., aquifer) may be problematic. The uses of evaporation ponds or ocean disposals are other potential options but will also impact the siting decision.

As mentioned above, the algae biofuels technology could utilize another source of low-quality water, co-produced water, associated with oil and natural gas extraction. Unfortunately, data on co-produced water volumes and disposal methods in India was not available. It is possible that this information exists but was not obtained due to access constraints given the proprietary nature of this type of information. Figure 13 shows the geographic distribution of oil and natural gas fields in India to illustrate existing locations of produced water resources. This information could be used to guide future, more refined analysis related to this resource potential.

Production of crude oil in India is about 34 Mt annually or approximately 252 million barrels (bbl)[2] (MPNG 2009). The global average water cut (i.e., the ratio of water produced from a well compared to the volume of total liquids) is estimated at about 75% (The Oil Drum 2009). This means that about 3/4 of the fluids brought up a well are water; therefore, this resource could be substantial in India. Natural gas wells typically produce much less water than oil wells (ANL 2009).

As shown in Figure 13, oil and natural gas in India are produced both onshore and offshore. Without information on the disposal method practiced in the country, it is difficult to estimate the amount of produced water that could be available for algae production. For offshore production activities, produced water is usually disposed of through direct ocean discharge after treatment, although sometimes it is transported to shore for disposal. Most of the production in India (about 70%) comes from offshore fields (MPNG 2009); therefore, it is possible that a large portion of the produced water in India is currently disposed of into the ocean. At onshore fields in many countries, most produced water is re-injected to maintain reservoir pressure and hydraulically drive oil toward a producing well (ANL 2009). A small portion of onshore-produced water is disposed of in onsite evaporation ponds or offsite treatment facilities, and some is used in agricultural and industrial activities if certain water quality conditions are met. The challenge to using produced water in the agricultural sector is its salinity, which is often greater than that of seawater. Moderately-saline produced water, up to 3,000 mg/L TDS, has been used for irrigation, but its continuous use, particularly in arid climates, leads to build-up of salts in the soil surface creating undesirable ecological conditions. The algae biofuels technology could take advantage of this poor-quality water by growing salt-tolerant species, thus avoiding pressure on freshwater resources. In addition to salt, however, there are other contaminants in produced water (oil and grease, inorganic and organic compounds, and radioactive material) that prevent its direct use and mandate treatment. These constituents could be harmful to algae as well. Detailed onsite evaluation of produced water resources would determine their potential for algae production.

Wastewater discharged from domestic, industrial, and agricultural activities is another source of low-quality water that could be used for algae production. Using wastewater effluent as pond medium in co-located algae farms with facilities treating sewage, industrial wastewater, and agricultural drainage waters provides a cost-effective solution not only to water but also to land and nutrients considerations because the wastewater treatment function would cover nearly all costs. There is also an environmental benefit of using the wastewater effluent for growing algae. These waters, particularly sewage effluent and agricultural runoff, are high in nutrients such as nitrate and phosphates and when released into rivers, lakes, or the ocean causes eutrophication— nutrient pollution—leading to negative impacts on water quality and aquatic ecosystems. Algae requires these nutrients to grow, as discussed in the next section of this report; therefore, using the wastewater effluent as an algae pond medium provides an effective way for nutrient removal and prevention of natural waters' pollution while creating a product with an economic value.

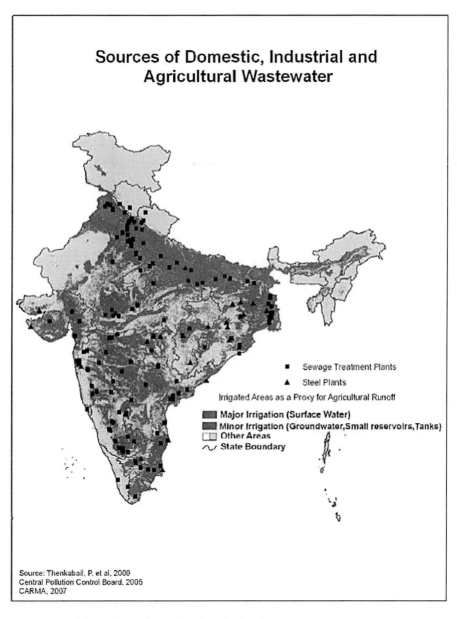

Figure 14. Sources of domestic, industrial, and agricultural wastewater.

According to the Central Pollution Control Board (CPCB 2005), there are about 269 sewage treatment plants (STP) in India, of which 231 are operational and 38 are under construction, located in 160 cities and towns. Given that the majority of the population in the country lives in rural areas (about 70%) and the STP are located in urban areas, it is evident that sewage is not treated in rural areas and it is disposed of in nearby rivers, lakes, or the ocean. Perhaps the development of algae production facilities in rural areas coupled with wastewater treatment would stimulate not only economic growth in these areas but would also improve their environmental footprint. Similar benefits could be provided by using agricultural drainage waters for growing algae. Agricultural drainage water is the excess water that runs off the field at the low end of furrows, border strips, basins, and flooded areas during irrigation or rainfall. In some parts of the country, where water resources are scarce, drainage water is re-used to meet crop water requirements. This, however, is applicable to drainage water with relatively good quality. Often, in addition to high nutrients, drainage water is high in salt content or the salinity increases with each re-use, which prevents its use for irrigation, but it could be suitable for cultivating salt- tolerant algae species.

Sources of industrial wastewater include the iron and steel-manufacturing sector, mines and quarries, chemicals, and the food industries. It is estimated that a large quantity of industrial wastewater is generated in India: about 1,057 trillion cubic meters (Tm^3) in 1997 (Ravindranath et al. 2005). Most of this volume comes from steel plants (1,040 Tm^3), followed by the pulp and paper plants (7.2 Tm^3), distilleries (6 Tm^3), cotton plants (1.5 Tm^3), and other food-processing and chemicals facilities.

Figure 14 shows the location of STP and steel plants in India as well as the irrigated lands as a proxy for agricultural drainage waters. These locations should be further investigated as potential sites for co-locating algae farms, thus taking advantage of the existing infrastructure and providing ecological benefits to the area.

Carbon Dioxide and Other Nutrients

Optimal algae growth occurs in a CO_2-saturated environment. Therefore, algae production provides an excellent opportunity for the utilization of carbon emissions and serves as a complement to subsurface sequestration. India was the fourth-largest emitter of CO_2 in 2006, releasing 1,510 Mt into the atmosphere, or about 5% of the world total (United Nations 2009). The largest anthropogenic source of CO_2 emissions in India is the combustion of fossil fuels used in power generation, industrial processes, transportation, agriculture, commercial services, and the residential sector. It is expected that India will continue to expand coal power generation given its large coal reserves and increasing energy consumption. It is also expected that the transportation sector will continue to grow as vehicle ownership increases. Therefore, the country's continuous economic development, under a business-as-usual scenario, will lead to more CO_2 emissions in the future. The Indian Government projects that the country's CO_2 emissions will grow three to five times by 2031 (YSFES 2009).

Not all CO_2 emissions are suitable for capture. This is applicable to large stationary sources with high concentrations of CO_2. In India, these include thermal power plants, steel plants, cement plants, fertilizer plants, refineries, and petrochemical plants. Together, these facilities emitted 638 Mt of CO_2 in 2007, nearly half of the total CO_2 emissions in India (CARMA 2007). Figure 15 depicts the location of large stationary sources of CO_2 in India and their emissions in 2007. These facilities are widely distributed throughout the country with some large clusters providing an opportunity for large-scale algae production. The large clusters overlap with India's major industrial regions: Delhi, Mumbai, Ahmadabad, Kolkata, Madras, and Varanasi. There are other smaller clusters of stationary CO_2 sources in central India and around cities in other parts of the country.

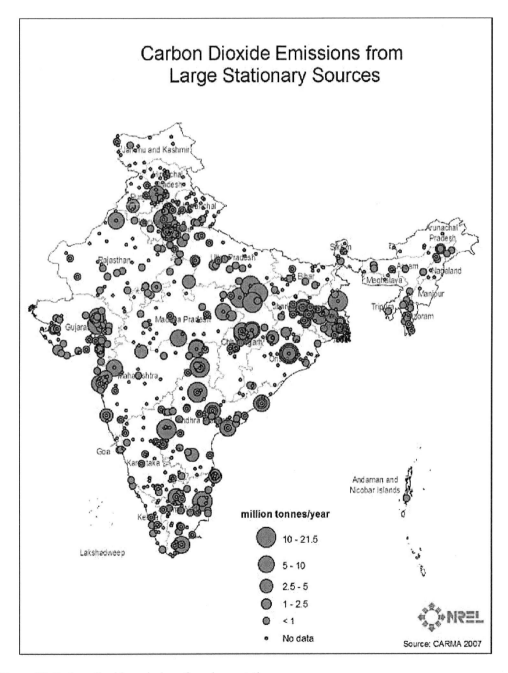

Figure 15. Carbon dioxide emissions from large stationary sources.

The concept of co-locating large point sources of CO_2 with algae farming provides an effective approach to recycle the CO_2 into a useable liquid fuel. To put this into perspective, capturing around 20% of the 638 Mt of CO_2 released into the atmosphere from large stationary sources by algae for conversion to fuels would be enough to displace 30% of diesel fuel currently used in India[3]. This is based upon an estimated 40 gallons of algal oil (about 0.13 tonnes) per tonne of CO_2 consumed during algal biomass production at 30% lipid algal content by weight (USDOE 2009). Therefore, while it will not be practical to assume that algal production could absorb all CO_2 emissions from India's large stationary sources, the CO_2 resources available can yield large quantities of algal oils and ultimately transportation fuels.

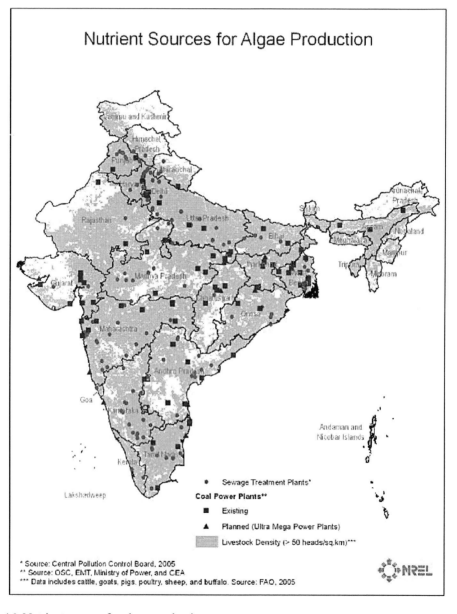

Figure 16. Nutrient sources for algae production.

In addition to CO_2, algae require nitrogen, phosphorous, sulfur, and other trace nutrients to grow; diatoms also require silicon for construction of the cell wall. For co-located algae farms with wastewater treatment facilities, the nitrogen and phosphorus would be provided by the effluent used as pond medium. For dedicated algae farms, sludge (the residual material left from the treatment process) could be used as well as animal manure, another nutrient-rich resource. It is important that dedicated algae farms be located near the nutrient sources to keep transportation costs low. Nitrogen could also be supplied by coal-burning power plants as nitrogen oxide (NOx). Additionally, coal-burning power plants could provide sulfur as sulfur oxide (SOx). Silicon for diatom growth will likely require supplementation. Figure 16 shows the distribution of nutrient sources in India to illustrate areas with opportunities for co-locating algae farms or situating them nearby. It must be noted that nutrient requirements, like growth temperature, must be maintained within an optimal range to promote maximal growth. Too little of any one particular nutrient will reduce the growth rate and conversely, too much of a nutrient can prove toxic. Nutrient limitation can result in increased overall lipid content in algal cells, but it comes at the expense of overall productivity (USDOE 2009).

Land

India has a total land area of approximately 297 million hectares (Mha)[4] . The share of agricultural land in 2007 was significant, about 60% (180 Mha) of the total land area (FAOSTAT 2007). The majority of this land is under cultivation for food crops, and only about 3.5% (10.4 Mha) of the total land area is dedicated to pasture. Forest land covers about 22% of the country's area, and the remaining 14.5% is other land (e.g., build-up areas and barren land).

The availability of land for algae production in India will depend on many physical, economic, legal, social, and political factors. Physical characteristics, such as topography and soil, could limit the land available for algae farming in some locations. Topography would be a limiting factor for these systems since the installation of large shallow ponds requires relatively flat terrain. Areas with more than a 5% slope[5] can be effectively eliminated from consideration due to the increased costs of site development. Soils, and particularly their porosity/permeability characteristics, affect the construction costs and design of open systems by virtue of the need for pond lining or sealing. Soils with low permeability (such as clay soils) are good for aquaculture as the water loss through seepage or infiltration is low. Figure 17 illustrates areas with flat terrain in India as well as locations where they overlap with clay soils. It shows that the majority of the country is suitable for algae production from topography's perspective except the mountain areas in the north, east, and some internal parts of the country. Large portions of central and eastern states such as Madhya Pradesh Maharashtra, Karnataka, and Gujarat have a combination of flat terrain and clay soils. Clay soils suggest challenges for food crop production but are the preferred soil type for installation of algae ponds; thus, they may present an opportunity for algae farming in areas where they exist. A more refined examination of these soils, including spatial distribution and land use, would determine the possibility of siting algae farms in the areas outlined in Figure 17.

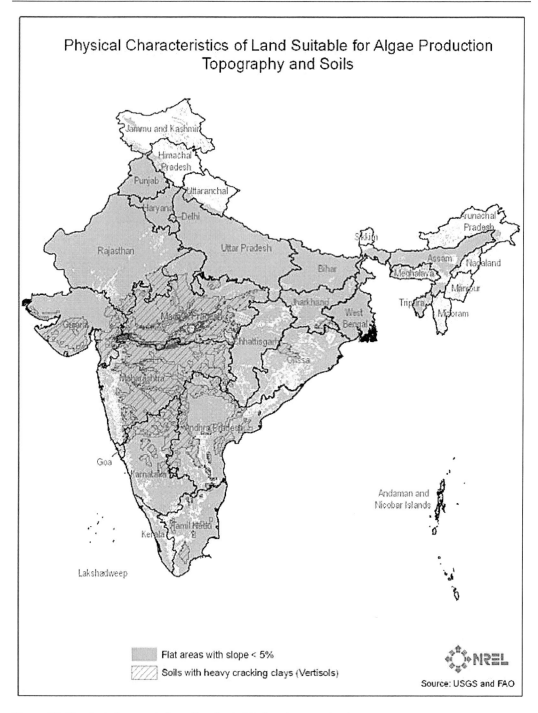

Figure 17. Physical characteristics of land suitable for algae production.

Land use is a very important consideration in evaluating locations suitable for algae farming, and certain land use categories will limit the land available to this technology. These categories include agricultural and forest lands, protected areas and other environmentally sensitive areas (such as biodiversity hot spots), and lands with cultural and historic value. Agricultural land in India covers a large portion of the country, as illustrated in Figure 18.

Therefore, to avoid competition with food/feed/livestock production, algae technology should focus on the degraded portions of these lands, called "wastelands" in India. The wastelands include degraded cropland and pasture/grazing land as well as degraded forest, industrial/mining wastelands, and sandy/rocky/bare areas. The Ministry of Rural Development's Department of Land Resources and the National Remote Sensing Agency (MRD-NRSA 2005) developed a very detailed assessment of wastelands in India. It estimated that the extent of these lands is about 55 Mha (approximately 17% of total land area). States with the largest amount of wasteland include Rajasthan, Jammu and Kashmir, Madhya Pradesh, Maharashtra, and Andhra Pradesh. The results of this study are shown in Table 1, and the Appendix contains state-specific maps illustrating the spatial distribution of these lands (Figures 28–59). This information would be very useful to policymakers and industry developers in selecting the most suitable locations for algae farming.

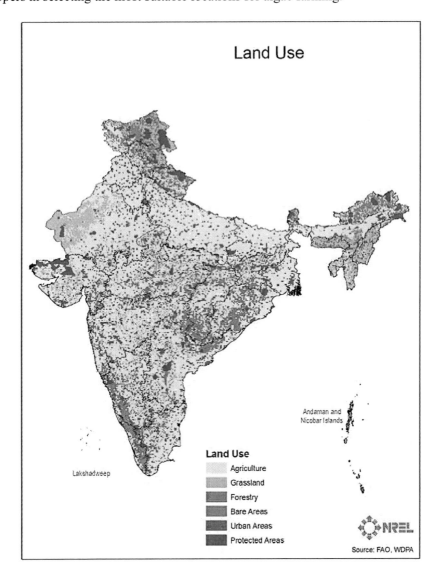

Figure 18. Land use.

Table 1. Wasteland in India by State

State	Total Area (ha)	Wasteland	
		ha	%
Andaman and Nicobar	824,900	N/A	N/A
Andhra Pradesh	27,506,800	4,526,700	16.46
Arunachal Pradesh	8,374,300	1,817,600	21.70
Assam	7,843,800	1,403,400	17.89
Bihar	9,417,100	544,400	5.78
Chandigarh	14,400	N/A	N/A
Chhattisgarh	13,519,400	758,400	5.61
Dadra and Nagar Haveli	49,100	N/A	N/A
Daman and Diu	12,200	N/A	N/A
Delhi	148,300	N/A	N/A
Goa	370,200	53,100	14.34
Gujarat	19,602,400	2,037,800	10.40
Haryana	4,421,200	326,600	7.39
Himachal Pradesh	5,567,300	2,833,700	50.90
Jammu and Kashmir*	22,223,600	7,020,200	31.59
Jharkhand	7,970,600	1,116,500	14.01
Karnataka	19,179,100	1,353,700	7.06
Kerala	3,886,300	178,900	4.60
Lakshadweep	3,200	N/A	N/A
Madhya Pradesh	30,825,200	5,713,400	18.53
Maharashtra	30,769,000	4,927,500	16.01
Manipur	2,232,700	1,317,500	59.01
Meghalaya	2,242,900	341,100	15.21
Mizoram	2,108,100	447,000	21.20
Nagaland	1,657,900	370,900	22.37
Orissa	15,570,700	1,895,300	12.17
Puducherry	49,200	N/A	N/A
Punjab	5,036,200	117,300	2.33
Rajasthan	34,223,900	10,145,400	29.64
Sikkim	709,600	380,800	53.66
Tamil Nadu	13,005,800	1,730,300	13.30
Tripura	1,048,600	132,300	12.62
Uttar Pradesh	24,092,800	1,698,400	7.05
Uttaranchal	5,348,300	1,609,700	30.10
West Bengal	8,875,200	439,800	4.96
Total	328,730,300	55,237,700	16.80

Source: MRD–NRSA 2005; * Area not surveyed: 12,084,900 ha.

Other considerations that may limit the land available for algae production are land ownership and land price. Land ownership information provides valuable insights on which policies and parties could affect project development. In India, most of the agricultural land is privately owned and some is under the jurisdiction of state governments. State governments often are the middleman in land acquisition and distribution. The majority of farm holdings are small—the average size of operational farms is about 1.18 ha (New Agriculturist 1999). Therefore, acquiring land for large-scale production of algae may present some challenges considering that the process may have to involve many parties; although, land takeover in India is not considered a problem when adequate compensation is offered.

Data on land prices in India is difficult to obtain due to the high variability between and within the states. Land prices are hard to generalize because they depend on numerous factors including location, availability of water, and proximity to transportation infrastructure. Unfortunately, no centralized agency that collects and disseminates this type of information was identified. Given that the algae technology's strategy is to use wastelands and low-quality water, it is expected that the price of those lands would be favorable.

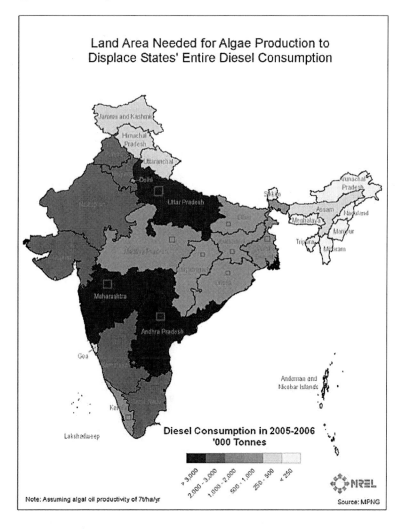

Figure 19. Land area needed for algae production to displace states' entire diesel consumption.

Table 2. Land Area Needed for Algae Production to Displace States' Entire Diesel Consumption

State	Total Land Area (ha)	Diesel Consumption in 2005-06 (tonnes)*	Area Needed for Algae Production to Displace States' Entire Diesel Consumption	
			ha	%
Andaman and Nicobar	824,900	73,571	10,510	1.27
Andhra Pradesh	27,506,800	3,232,424	461,775	1.68
Arunachal Pradesh	8,374,300	44,682	6,383	0.08
Assam	7,843,800	446,960	63,851	0.81
Bihar	9,417,100	921,393	131,628	1.40
Chandigarh	14,400	63,578	9,083	63.07
Chhattisgarh	13,519,400	610,297	87,185	0.64
Dadra and Nagar Haveli	49,100	60,403	8,629	17.57
Daman and Diu	12,200	52,055	7,436	60.95
Delhi	148,300	1,163,193	166,170	112.05
Goa	370,200	272,535	38,934	10.52
Gujarat	19,602,400	2,056,946	293,849	1.50
Haryana	4,421,200	2,306,747	329,535	7.45
Himachal Pradesh	5,567,300	306,835	43,834	0.79
Jammu and Kashmir	22,223,600	365,778	52,254	0.24
Jharkhand	7,970,600	889,122	127,017	1.59
Karnataka	19,179,100	2,387,767	341,110	1.78
Kerala	3,886,300	1,333,875	190,554	4.90
Lakshadweep	3,200	4,323	618	19.30
Madhya Pradesh	30,825,200	1,380,531	197,219	0.64
Maharashtra	30,769,000	3,345,770	477,967	1.55
Manipur	2,232,700	31,939	4,563	0.20
Meghalaya	2,242,900	165,090	23,584	1.05
Mizoram	2,108,100	25,825	3,689	0.18
Nagaland	1,657,900	27,045	3,864	0.23
Orissa	15,570,700	966,823	138,118	0.89
Puducherry	49,200	257,920	36,846	74.89
Punjab	5,036,200	2,110,946	301,564	5.99
Rajasthan	34,223,900	2,412,911	344,702	1.01
Sikkim	709,600	33,293	4,756	0.67
Tamil Nadu	13,005,800	2,981,109	425,873	3.27
Tripura	1,048,600	42,941	6,134	0.59
Uttar Pradesh	24,092,800	4,014,981	573,569	2.38
Uttaranchal	5,348,300	324,086	46,298	0.87
West Bengal	8,875,200	1,593,997	227,714	2.57
Total	328,730,300	36,307,691	5,186,813	1.58

*Source: MPNG 2009.

A rather large surface area would be required for commercial-scale algae production (either open ponds or PBRs). This is due to the fact that increasing culture depth does not increase productivity per unit area because solar radiation cannot penetrate deep into the culture. In other words, light is ultimately the limiting resource. Algae productivity is measured in terms of biomass produced per day per unit of horizontal surface area. Assuming a reasonably conservative average productivity of between 8 and $20g/m^2/day$ with 15% oil content on a dry weight basis, a hectare could produce about 4 to 10 tonnes (t) of algal oil (biodiesel) per year, which is double the value of the next highest producing crop, oil palm, which generates between 2 and 5 t/ha/yr. India's current diesel consumption is 52 Mt per year (MPNG 2009). Therefore, to substitute 20% of this consumption, algae production would need about 1–2.6 Mha while *Jatropha curcas,* another feedstock considered for biodiesel production in the country, with best yield of 1 t/ha/yr would need 10.4 Mha of land (most likely requiring irrigation and fertilizer). Moreover, an aggressive R&D program has the potential to reach much higher productivity— theoretical analyses indicate that productivity of over 37 t/ha/yr are possible (Weyer et al. 2009). Therefore, the land needed for algae production to substitute 20% of current diesel consumption would be much less, about 280,000 ha, and approximately 1.4 Mha would displace the country's entire diesel consumption. Figure 19 and Table 2 illustrate visually and statistically the diesel consumption by state and the land area needed for algae production to meet this demand (indicated by boxes in Figure 19) using an average near-term productivity rate of 7 t/ha/yr.

SITE-SUITABILITY ANALYSIS

This section builds upon the results of the resource evaluation conducted previously to identify areas in India suitable for algae farming. Two algae production facility types are considered: *co- located* facilities, which produce algal biomass as a consequence of another process, and *dedicated* facilities, with the main purpose of algae production. Co-located facilities are those operating in conjunction with wastewater treatment where algae are produced as a byproduct of the wastewater treatment process. In the near-term, algae production at wastewater treatment facilities is considered the "low-hanging fruit" of algal biofuels. The knowledge and experience gained at these facilities could provide the base, in the long-term, for stand-alone, dedicated biofuels production systems. As mentioned earlier, using the wastewater effluent as pond medium provides a cost-effective solution to water, land, and nutrients considerations because the wastewater treatment function would cover nearly all costs. Given the favorable climate conditions in India—considerable sunshine and generally warm weather—and simplicity of construction and operation, engineered waste stabilization ponds (WSP) are successfully operating in the country as one of the main wastewater treatment technologies. WSP are any pond or pond system designed for biological waste treatment. These ponds present an opportunity for early development of algae production systems in India. Microalgae, unlike terrestrial plants that capture CO_2 from the atmosphere, require concentrated sources of CO_2, such as power plants flue gas, for high productivity. Therefore, if algae farming is considered in conjunction with sewage wastewater treatment, additional CO_2 would be needed, preferably from a nearby source to keep the transportation costs low. Figure 20 shows the location of WSPs in India and the proximity of stationary CO_2 sources to illustrate opportunities for co-locating algae farms. It also shows the location of steel plants, a major source of industrial wastewater in the country. The steel industry, in addition to being capable of providing water, can also supply CO_2 for the cultivation of algae. If algae farming is considered in conjunction with steel plants, it most likely will require additional nutrients. The steel plants illustrated in Figure 20 are located in areas with a large concentration of livestock production, which could supply animal manure as algae fertilizer.

The analysis of potential locations for co-located algae facilities considers only domestic (WSP in particular, not all sewage treatment plants) and industrial wastewater treatment facilities (steel plants). Agricultural wastewater (such as drainage water) is not included due to lack of data; therefore, there are additional opportunities for co-locating algae farms in other parts of the country that should be further investigated.

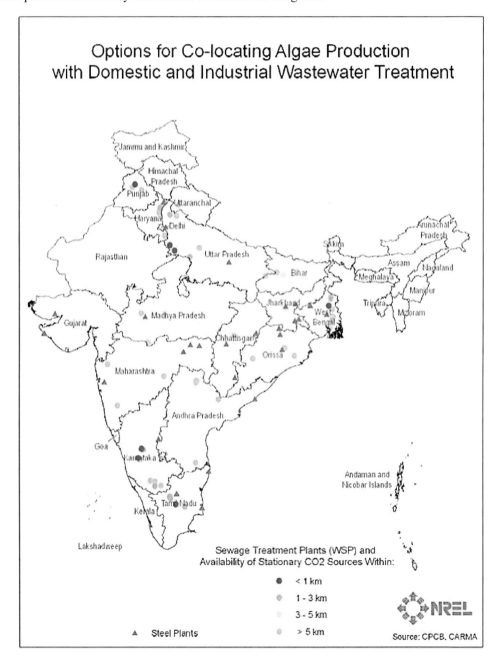

Figure 20. Options for co-locating algae production with domestic and industrial wastewater treatment.

Resource Evaluation and Site Selection for Microalgae Production in India

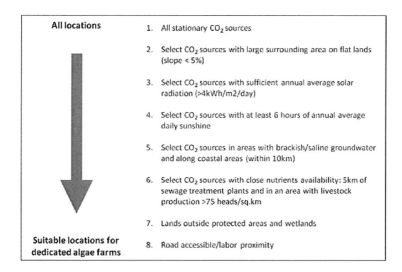

Figure 21. Methodology for selecting suitable locations for dedicated algae farms.

To identify suitable locations for dedicated algae farms in India, this study applies a series of steps as outlined in Figure 21 using GIS techniques. The analysis is centered on the stationary industrial CO_2 sources given key advantages of co-locating algae production with these facilities: supplying carbon for enhanced algal growth with low/no transportation costs and a means for capturing CO_2 before it is released to the atmosphere, thus providing potential carbon credits for utilities and industrial facilities. This combination of enhanced algal growth and GHG reduction makes co-location of algal cultivation with industrial CO_2 sources a promising option for future dedicated algae farms in India.

The analysis begins with the location of stationary CO_2 sources in India and looks for areas where these facilities coincide with other inputs necessary for algae growth or conditions that meet the engineering, economic, environmental, and social requirements for this technology.

These include flat terrain, sufficient solar radiation and daily sunshine hours, and availability of low-quality water and nutrients. In addition, these locations and their surrounding areas need to be outside of protected areas and wetlands, road accessible, and in proximity to populated places (labor availability). The results of this analysis are illustrated in Figure 22. Based on the available information, the most suitable locations for algal cultivation facilities in India are in the western and southern parts of the country and along coastal areas.

The next step in the analysis, which is beyond the scope of this project, is to evaluate the land availability for algae farming around the suitable stationary CO_2 sources, preferably wastelands, to avoid competition with valuable fertile agricultural land. This evaluation would be able to determine whether or not the wastelands estimated by the MRD-NRSA (2005) are actually available for algae production because some of these lands could already be in use. This detailed level of information is not readily available; thus, the analysis should be performed on site. Also, the land availability would be determined by the size of potential algae farms. A general examination of the suitable stationary CO_2 sources (Figure 22) and wastelands in India (Appendix) suggests that most of these sources are surrounded by or in close proximity to wastelands. Further, more detailed analysis would reveal the potential for algae production on these lands.

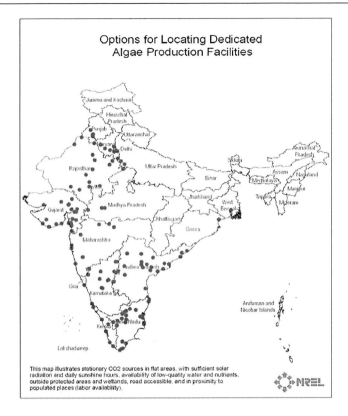

Figure 22. Options for locating dedicated algae production facilities.

CONCLUSIONS

The results of this study indicate that India has very favorable conditions to support algae farming for biofuels production. Most of the country meets the climate requirements for growing algae: sufficient solar radiation, sunshine hours, and warm temperatures. A large number of stationary CO_2 sources—thermal power plants, steel plants, cement plants, fertilizer plants, refineries, and petrochemical plants—are scattered throughout the country thereby providing opportunities for co-locating algae farms and recycling the CO_2 into a useable liquid fuel. Given the country's large concentration of population and agricultural activities, human and animal organic wastes are available as sources of nutrients such as nitrogen and phosphorous needed for algae growth.

Water resources in many parts of India are limited. Therefore, in order to be sustainable, the algae biofuels technology should not put additional demand on freshwater supplies and use low- quality water such as brackish/saline groundwater, co-produced water associated with oil and natural gas extraction, agricultural drainage waters, and other wastewaters. Although information on the quantity of these water resources in India is not available, the intensity of activities associated with their production suggests that there is a vast potential in the country.

Sustainable algae production also implies that the farming facilities would not be located on valuable fertile agricultural lands but on marginal lands (classified as wastelands in India).

These lands include degraded cropland and pasture/grazing land, degraded forest, industrial/mining wastelands, and sandy/rocky/bare areas. It is estimated that the extent of these lands in the country is about 55.27 Mha (approximately 18% of the total land area). If India dedicates only 10% (5.5 Mha) of its wasteland to algae production, it could yield between 22–55 Mt of algal oil[6] which would displace 45%–1 00% of current diesel consumption[7]. The production of this amount of algae would consume about 169–423 Mt of CO_2[8] which would offset 26%–67% of the current emissions[9].

This study illustrated options for siting two algae production facility types: co-located facilities, which produce algal biomass as a consequence of another process, and dedicated facilities, with the main purpose of algae production. Co-located facilities are those operating in conjunction with wastewater treatment where algae are produced as a byproduct of the wastewater treatment process. Using the wastewater effluent as pond medium provides a cost-effective solution to water, land, and nutrients considerations because the wastewater treatment function would cover nearly all costs. Domestic and industrial wastewater treatment facilities were considered in the analysis as potential sites for co-locating algae farms. This opportunity exists in most of the states in India.

The site-suitability analysis for dedicated algae farms was centered on the stationary industrial CO_2 sources given key advantages of co-locating algae production with these facilities: supplying carbon for enhanced algal growth with low/no transportation costs and a means for capturing CO_2 before it is released to the atmosphere, thus providing potential carbon credits for utilities. The study considered stationary CO_2 sources in areas where these facilities coincide with other inputs necessary for algae growth or conditions that meet the engineering, economic, environmental, and social requirements for this technology. The results of this analysis indicate that suitable locations for dedicated algae farms are in the western and southern parts of the country and along coastal areas.

It is important to note that the quality of the data used in this study is uncertain (most datasets were not checked for accuracy), and that due to lack of data, some resources were not evaluated (particularly co-produced water and agricultural wastewater). Therefore, the results of the site- suitability analysis are not complete and additional opportunities for co-located or dedicated algae farming systems most likely exist in other parts of the country. The purpose of this site- suitability analysis was to illustrate a methodology for refining the prospecting process of site identification and serve as a base for further analysis. This methodology could be revised to include other inputs if additional data are made available.

Future work could focus on validating the quality of existing data and gathering additional information such as the location of agricultural runoff collection sites, evaporative ponds used by the oil/gas production and mining industry, and other wastewater treatment facilities. It would also be helpful to collect information on the volume of wastewater. An analysis of the location and volume of low-quality water resources in India would be particularly useful in understanding better this resource potential not only for the production of algae biofuels but for other technologies as well.

India is a large country with diverse landscape and resources. Therefore, future work could focus on a state or even smaller geographic area to provide a more detailed examination of the resource potential for algae production and pinpoint the most suitable locations. The authors believe that the information provided in this study will serve as a base for further analysis of the algae biofuels potential in India and assists policymakers and industry developers in their strategic decisions.

APPENDIX

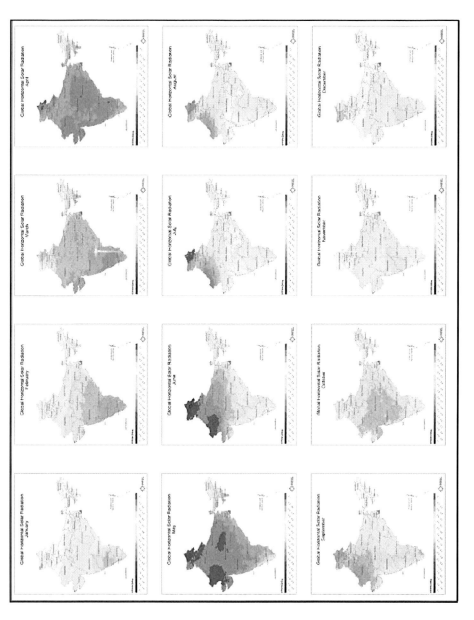

Figure 23. Monthly average solar radiation.

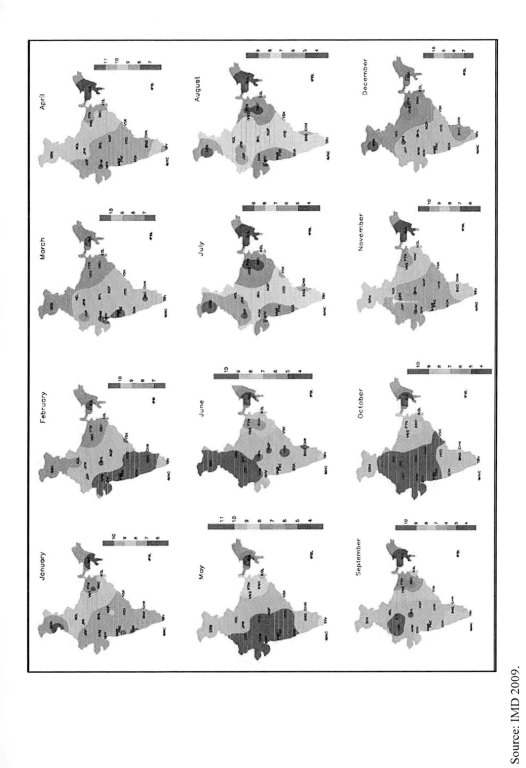

Source: IMD 2009.

Figure 24. Monthly average daily sunshine hours.

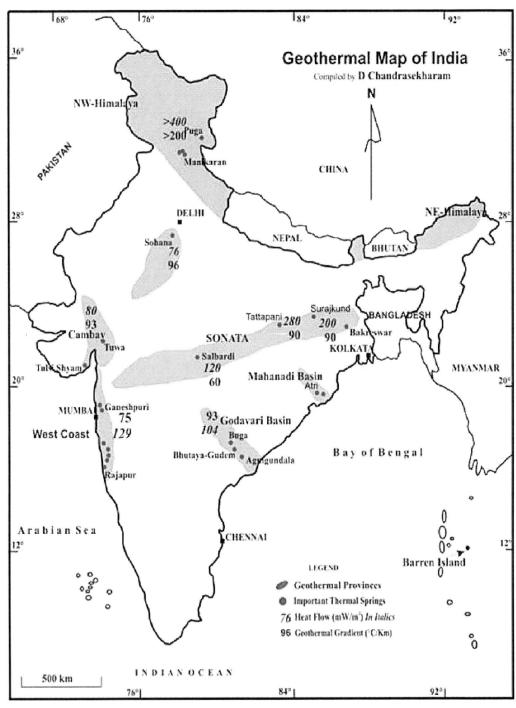

Source: Chandrasekharam, D.

Figure 25. Geothermal resources in India.

Resource Evaluation and Site Selection for Microalgae Production in India

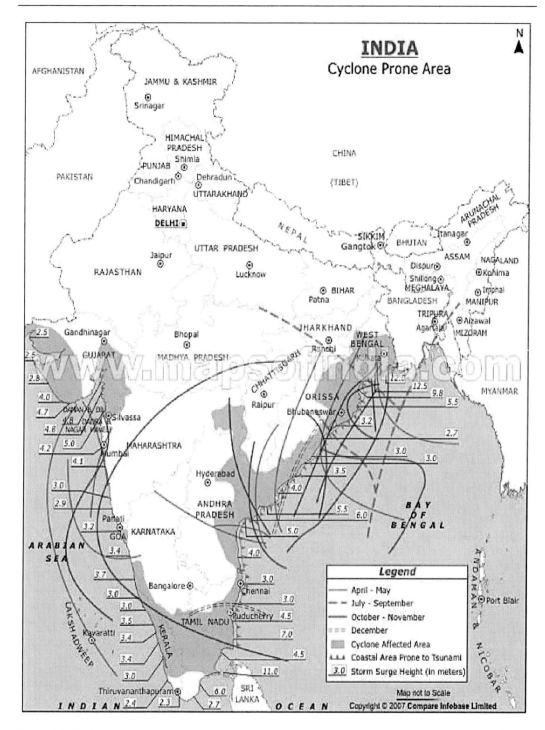

Figure 26. Cyclone prone areas.

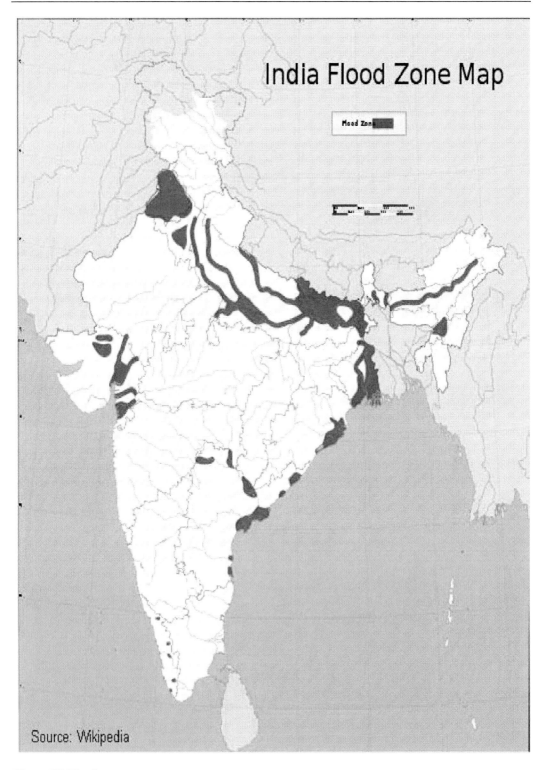

Figure 27. Flood zones.

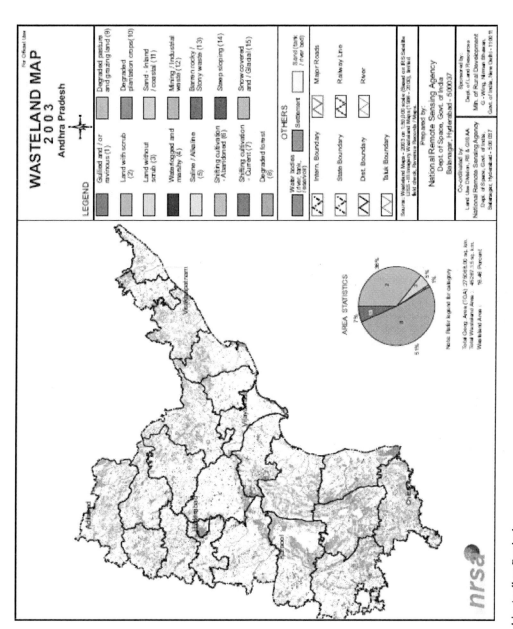

Figure 28. Wasteland in Andhra Pradesh.

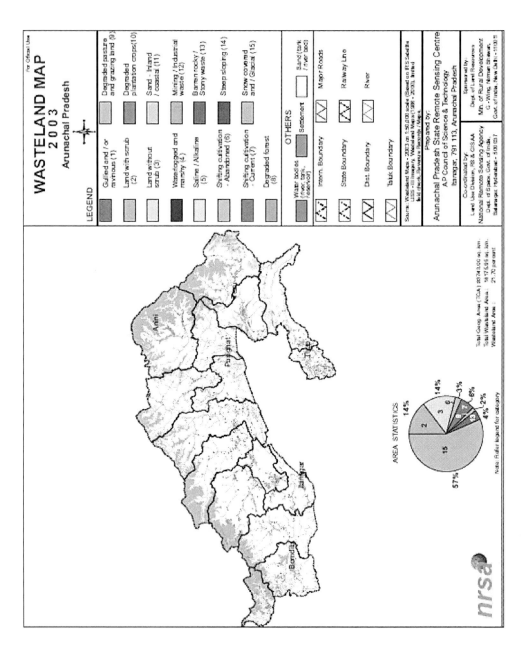

Figure 29. Wasteland in Arunachal Pradesh.

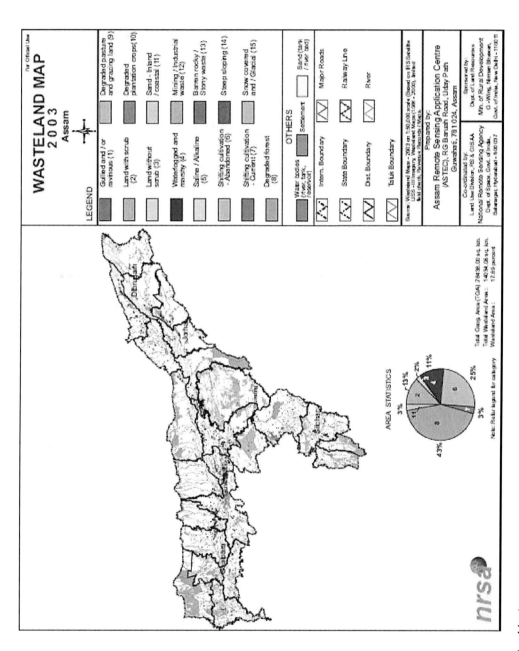

Figure 30. Wasteland in Assam.

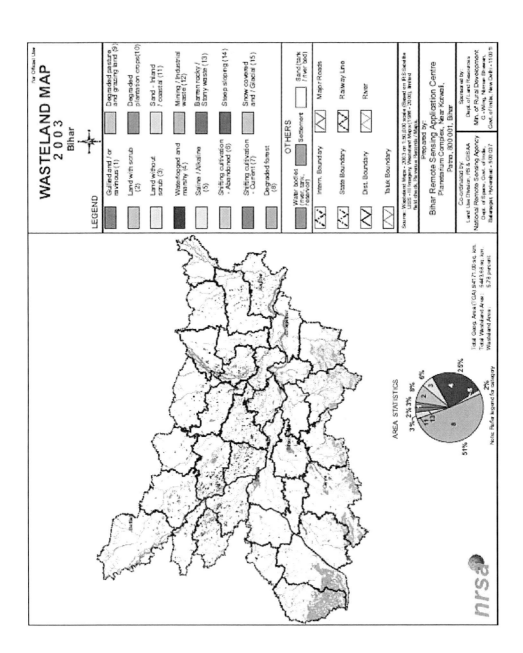

Figure 31. Wasteland in Bihar.

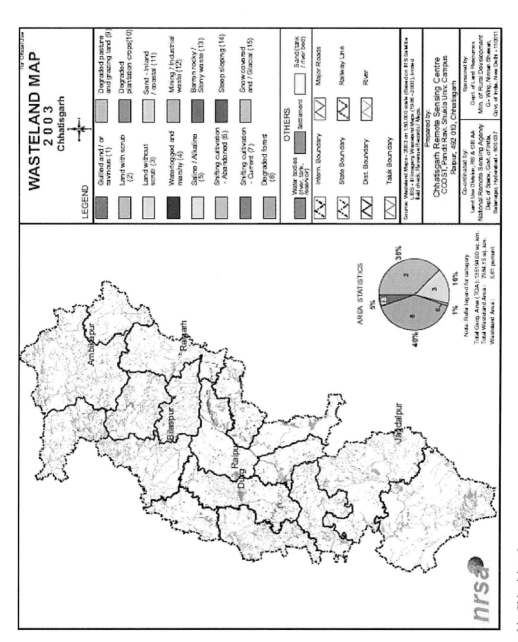

Figure 32. Wasteland in Chhatishgarh.

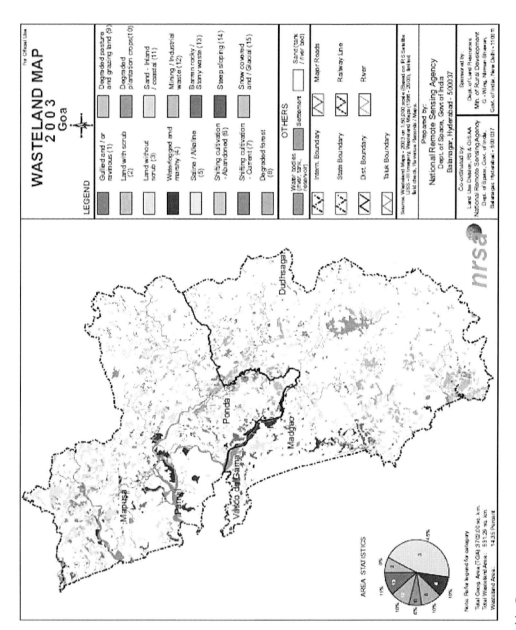

Figure 33. Wasteland in Goa.

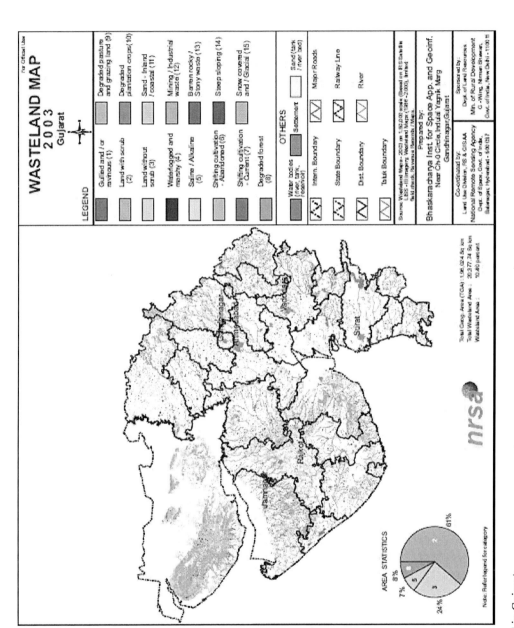

Figure 34. Wasteland in Gujarat.

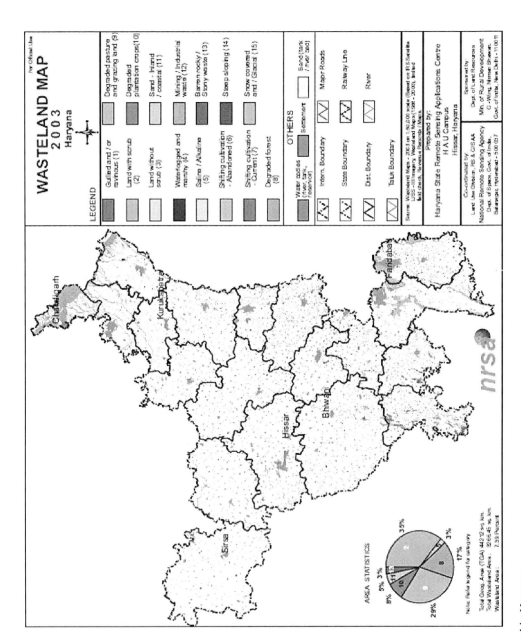

Figure 35. Wasteland in Haryana.

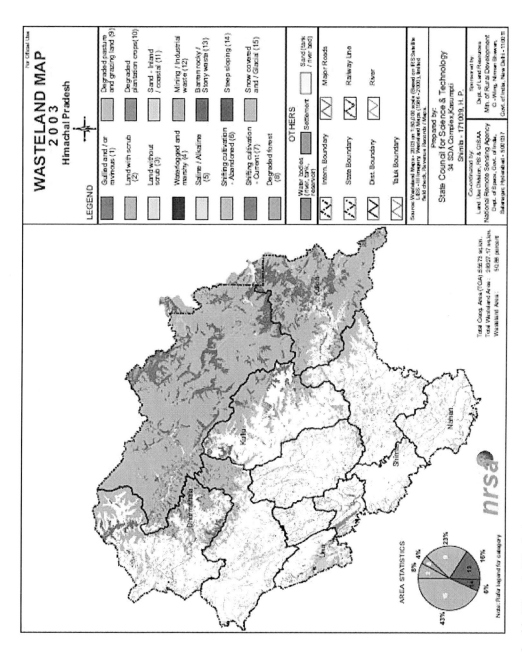

Figure 36. Wasteland in Himachal Pradesh.

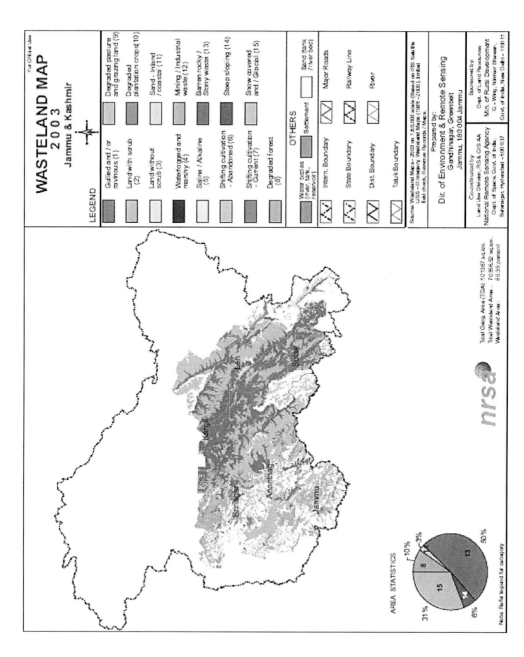

Figure 37. Wasteland in Jammu & Kashmir.

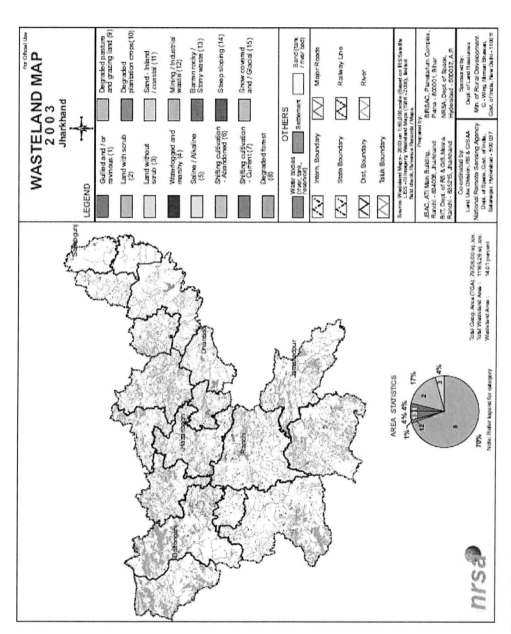

Figure 38. Wasteland in Jharkhand.

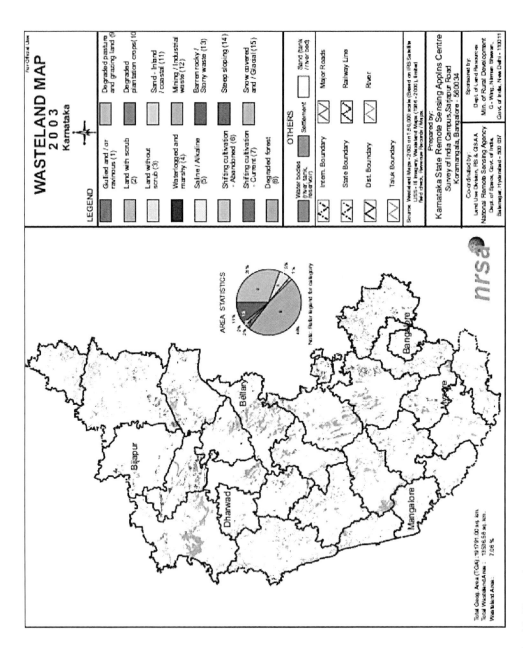

Figure 39. Wasteland in Karnataka.

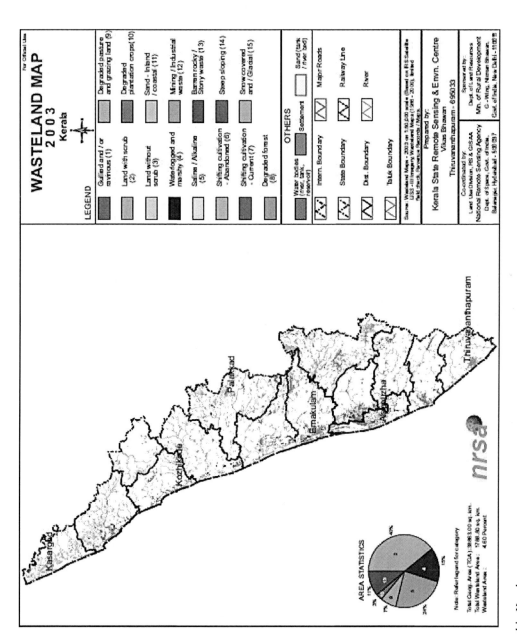

Figure 40. Wasteland in Kerala.

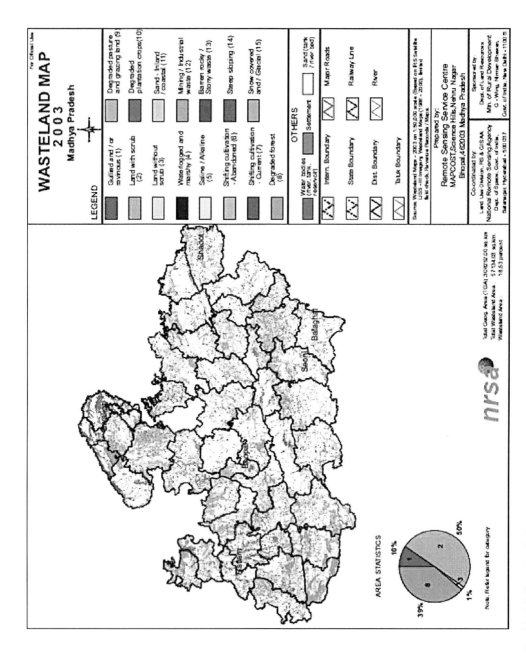

Figure 41. Wasteland in Madhya Pradesh.

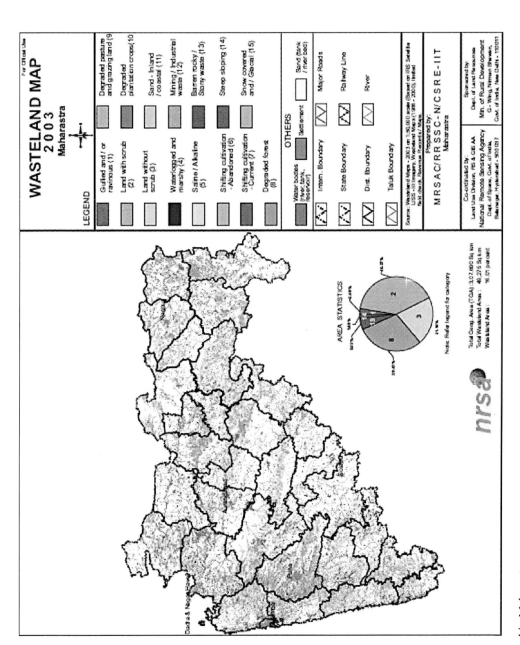

Figure 42. Wasteland in Maharastra.

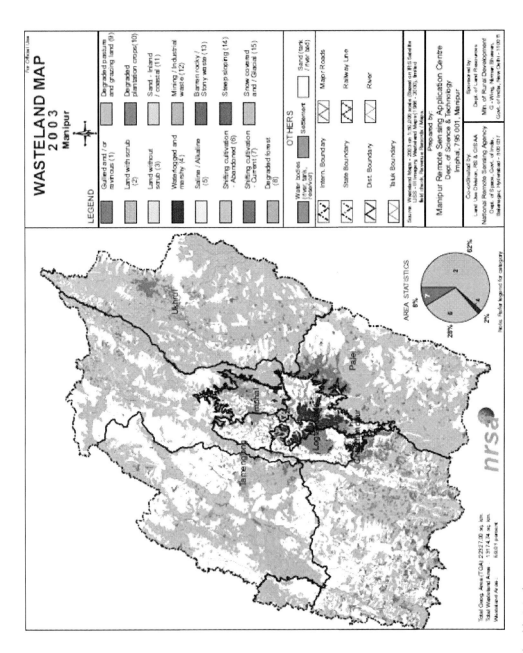

Figure 43. Wasteland in Manipur.

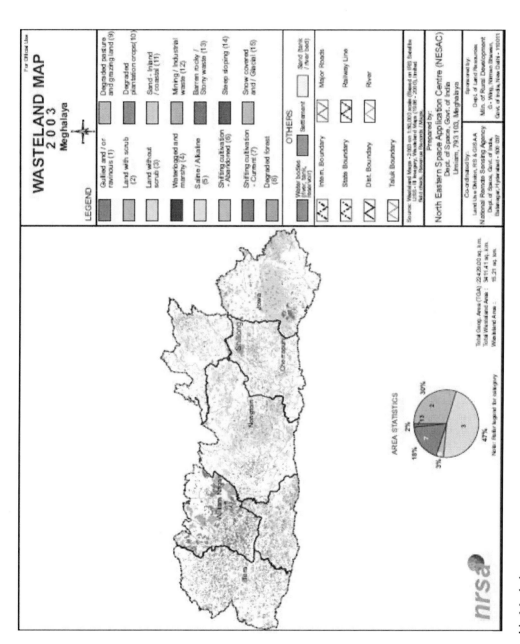

Figure 44. Wasteland in Meghalaya.

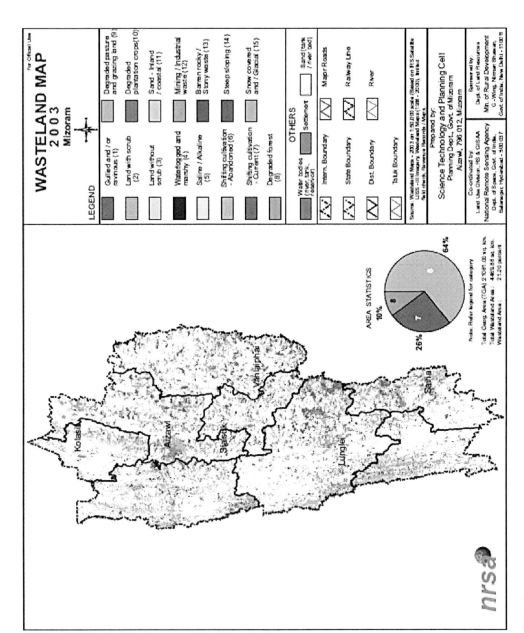

Figure 45. Wasteland in Mizoram.

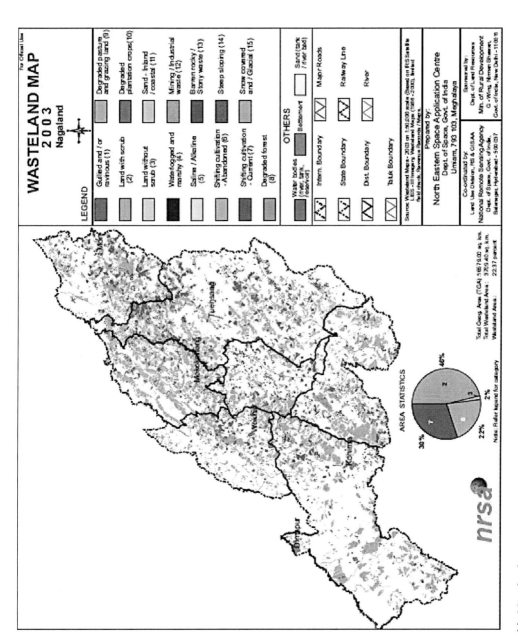

Figure 46. Wasteland in Nagaland.

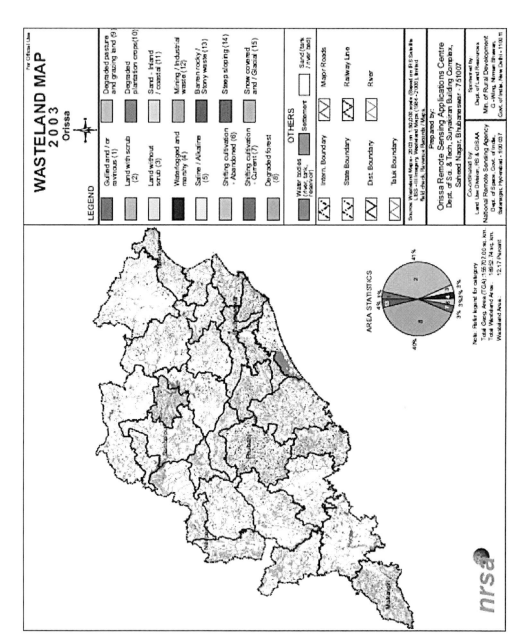

Figure 47. Wasteland in Orissa.

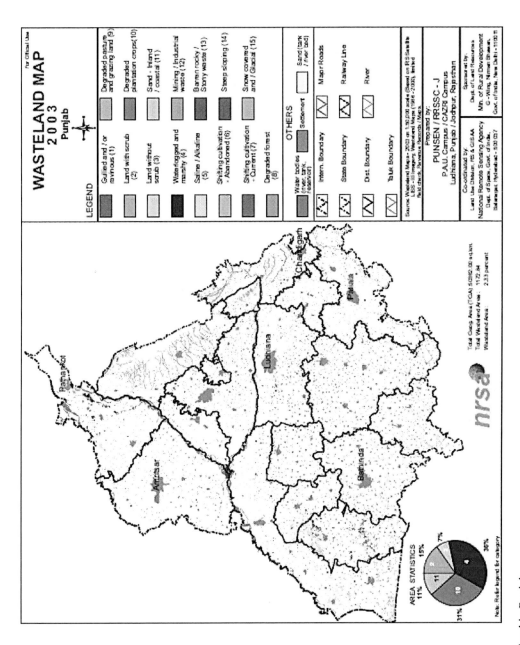

Figure 48. Wasteland in Punjab.

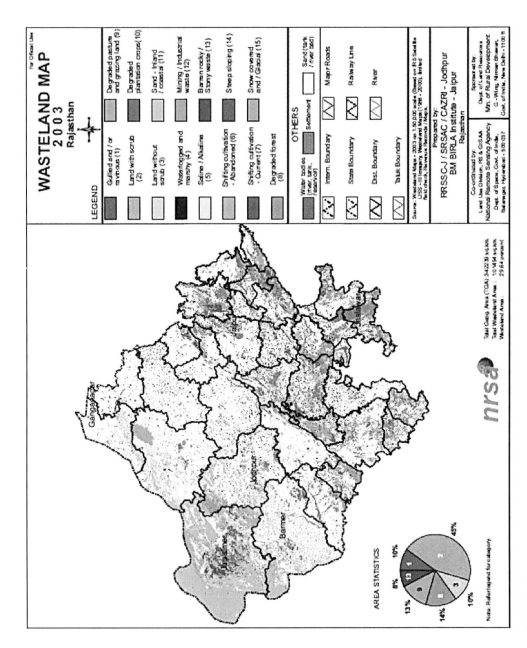

Figure 49. Wasteland in Rajasthan.

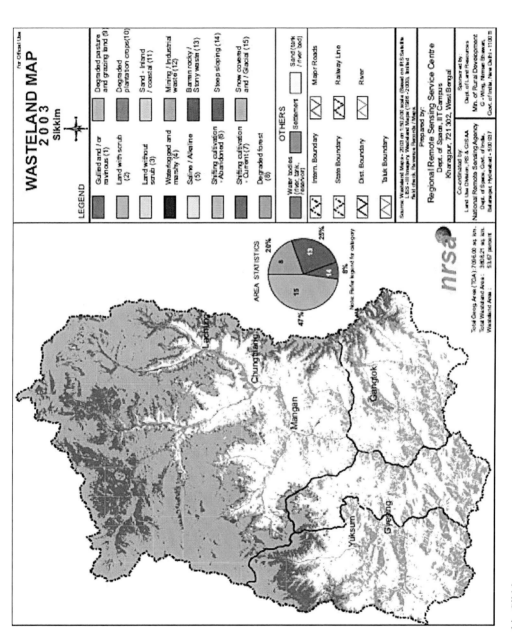

Figure 50. Wasteland in Sikkim.

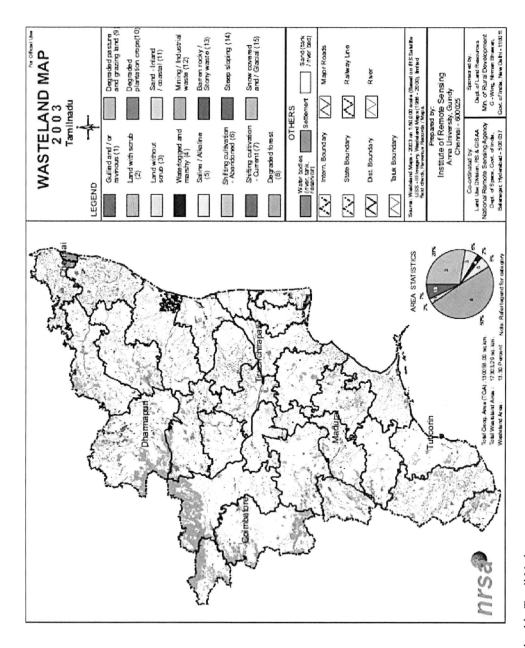

Figure 51. Wasteland in Tamil Nadu.

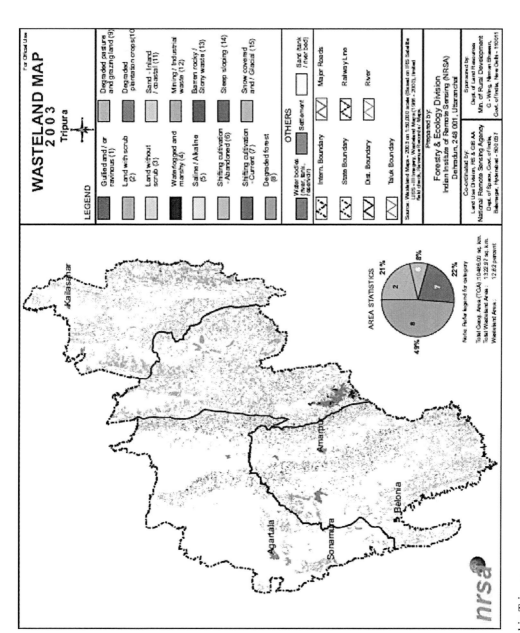

Figure 52. Wasteland in Tripura.

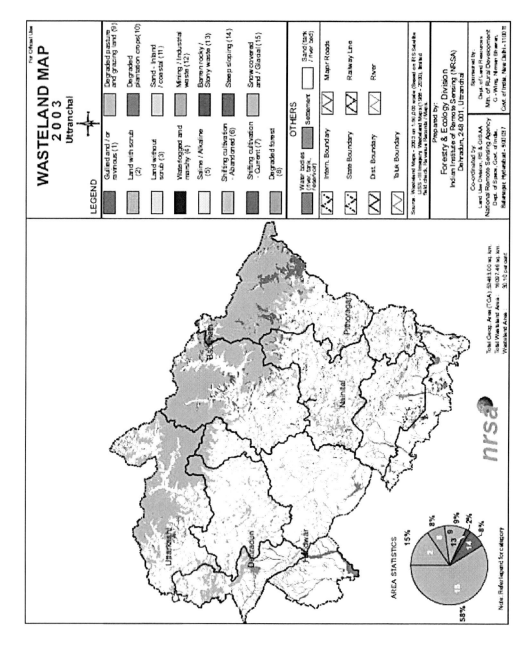

Figure 53. Wasteland in Uttaranchal.

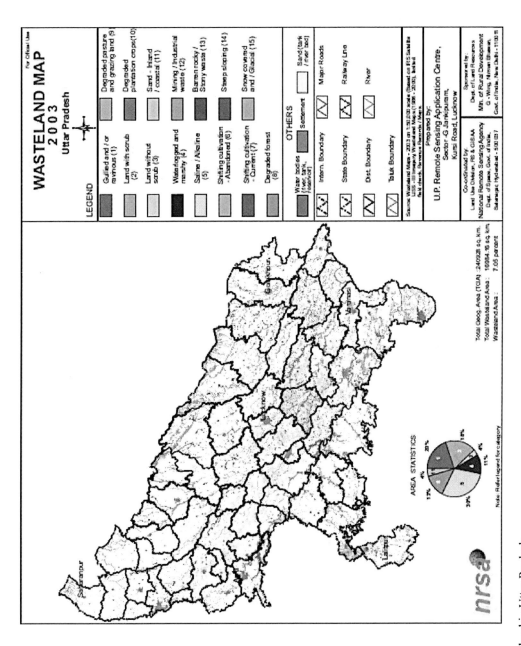

Figure 54. Wasteland in Uttar Pradesh.

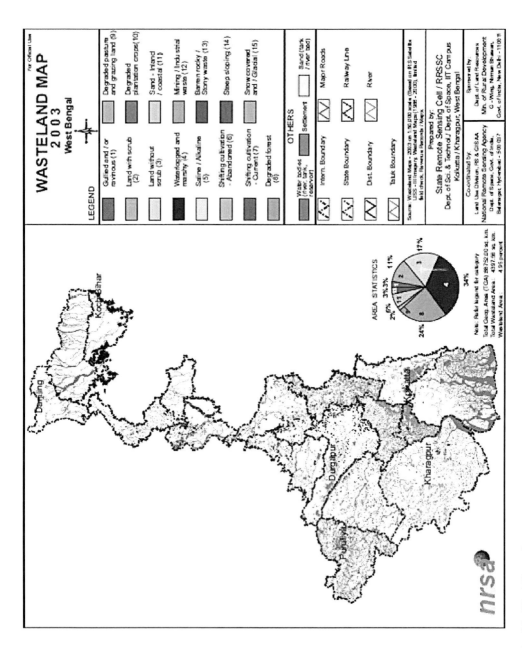

Figure 55. Wasteland in West Bengal.

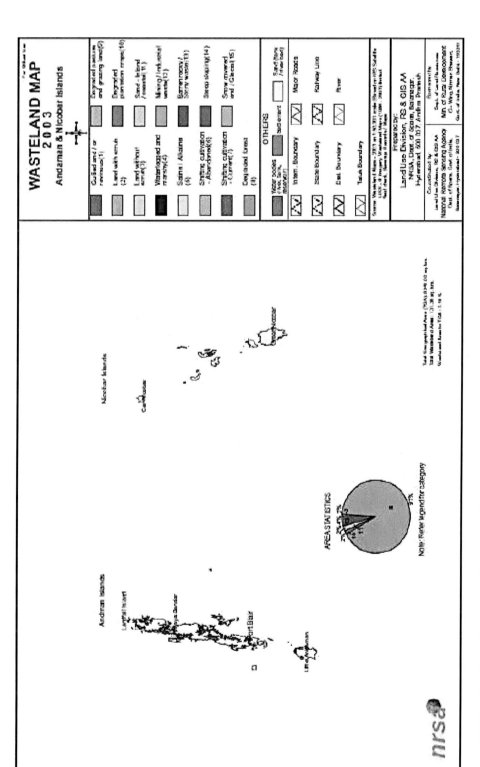

Figure 56. Wasteland in Andaman and Nicobar Islands.

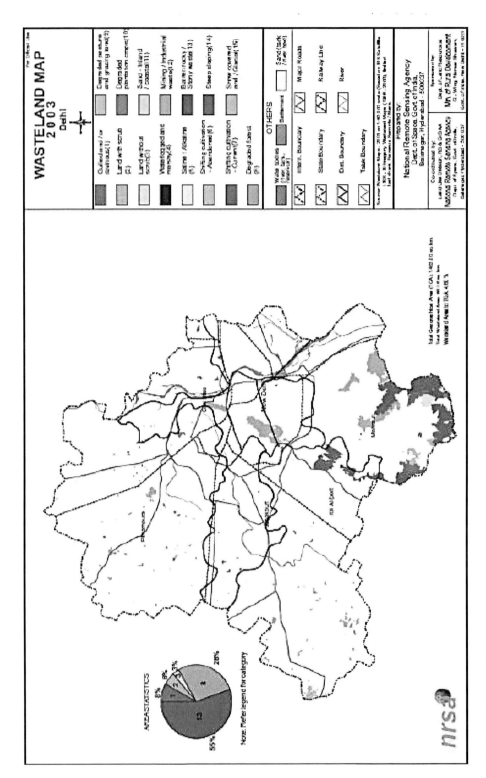

Figure 57. Wasteland in Delhi.

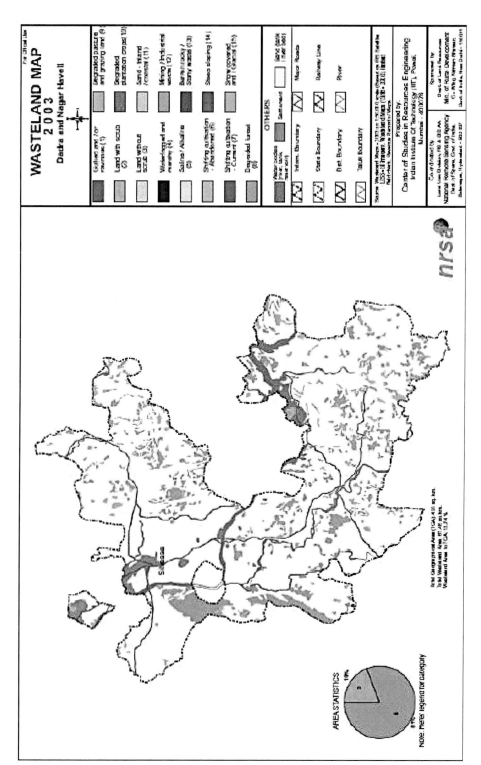

Figure 58. Wasteland in Dadra and Nagar Haveli.

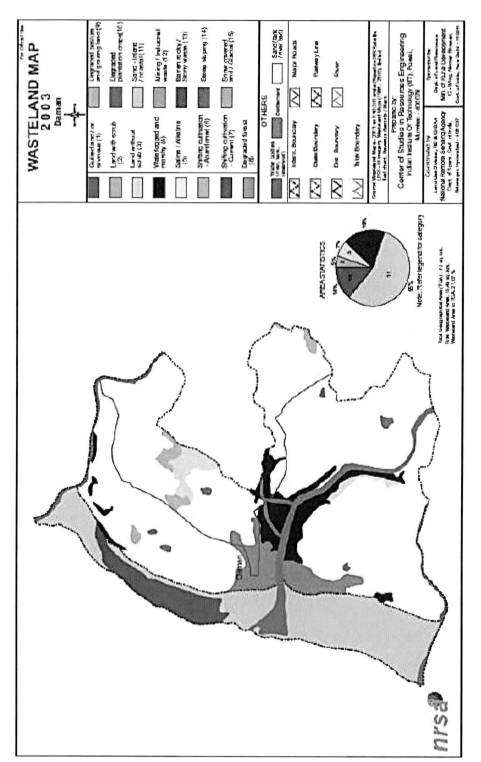

Figure 59. Wasteland in Daman.

REFERENCES

[1] Argonne National Laboratory (ANL). (September 2009). *"Produced Water Volumes and Management Practices in the United States."* http://www.ead.anl.gov/pub/doc/ANL EVS R09 produced water volume report 2437.pdf. Accessed May 27, 2010.

[2] Brown, L. (2007). *"Aquifer Depletion."* The Encyclopedia of Earth. http://www.eoearth.org/article/Aquifer depletion. Accessed May 27, 2010.

[3] Carbon Monitoring for Action (CARMA). (2007). http://carma.org/. Accessed December 2009.

[4] Central Ground Water Board (CGWB) & Ministry of Water Resources. (2006a). *"Dynamic Groundwater Resources of India."* http://cgwb.gov.in/GroundWater/GW assessment.htm. Accessed May 27, 2010.

[5] Central Ground Water Board (CGWB) & Ministry of Water Resources. (2006b). *"Groundwater Quality Monitoring."* http://cgwb.gov.in/GroundWater/Gw quality.html. Accessed November 2009.

[6] Central Ground Water Board (CGWB) & Ministry of Water Resources. (2006c). *"State Profile: Districts Affected by Salinity."* http://cgwb.gov.in/gw_profiles/st_ap.htm. Accessed November 2009.

[7] Central Intelligence Agency (CIA). (2009). *"Total Renewable Water Resources."* The World Factbook. https://www.cia.gov/library/publications/the-world-factbook/fields/2201.html. Accessed November 2009.

[8] Central Pollution Control Board (CPCB). (2005). *"Status of Sewage Treatment in India".* November. New Delhi, India.

[9] Chandrasekharam, D. (2009). *"Geothermal Energy Resources in India."* http://www.geosyndicate.com/dchandra/geoenergyresource.html. Accessed December 2009.

[10] Chelf, P. & Brown, L. M. (1989). *"Microalgae Mass Culture and the Greenhouse Effect: Resources Update."* Aquatic Species Program Annual Report (Bollmeier, W.S.; Sprague, S., eds.). SERI/SP-231-3579. Golden, CO: Solar Energy Research Institute, pp. 9-15.

[11] Chisti, Y. (2007). "Biodiesel from Microalgae." *Biotechnology Advances, (25)*, pp. 294-306.

[12] Directorate General of Hydrocarbons (DGH) & Ministry of Petroleum and Natural Gas. (2008). "Petroleum Exploration and Production Activities in India 2007-08." Noida, India.

[13] Eriksen, N. T. (2008) "The Technology of Microalgal Culturing." *Biotechnology Letters, (30)*, pp. 1525-1536.

[14] FAO. (2004). "Land and Livelihoods: Making Land Rights Real for India's Rural Poor." http://www.fao.org/docrep/007/j2602e/j2602e00.htm. Accessed May 27, 2010.

[15] FAOSTAT. (2007). "Land Use Database" http://faostat.fao.org/site/377/default.aspx#ancor. Accessed February 2010.

[16] Feinberg, D. A. & Karpuk, M. E. (1990). *CO_2 Sources for Microalgae Based Liquid Fuel Production.* SERI /TP-232-3820. Golden, CO: Solar Energy Research Institute.

[17] Harihara Ayyar, P. S. (1972). *"Water Resources of India in the Form of Rainfall."* India Meteorological Department. New Delhi-110003.

[18] Hindustan Times. (August 2005). *"Annual Average Rainfall During Last Five Years Estimated."* http://www.highbeam.com/doc/1P3-1093149621.html. Accessed May 27, 2010.

[19] Hu, Q., Sommerfield, M., Jarvis, E., Ghirardi, M., Posewitz, M. & Seibert, M., et al. (2008). "Microalgal Triacylglycerols as Feedstocks for Biofuel Production: Perspectives and Advances." *Plant Journal, (54)*, pp. 621–639.

[20] Huesemann, M. H. & Benemann, J. R. (2008). "Biofuels from Microalgae: Review of Products, Processes, and Potential, with Special Focus on Dunaliella Species." Ben-Amotz, A., Polle, J. E. W., Subba Rao, D. V., eds. *The Alga Dunaliella: Biodiversity, Physiology, Genomics, and Biotechnology,* New Hampshire: Science Publishers.

[21] India Meteorological Department (IMD). (2009). "Solar Radiant Energy Over India." New Delhi.

[22] Indian Institute of Tropical Meteorology Pune (via India Water Portal). (2009). "Physiographic Rainfall Variations Across India 1813-2006". http://www.indiawaterportal.org/image. Accessed October.

[23] International Energy Agency. (2007). "World Energy Outlook 2007: China and India Insights." http://www.iea.org/weo/2007.asp. Accessed May 27, 2010.

[24] International Energy Agency & Greenhouse Gas Programme. (May 2008). *A Regional Assessment of the Potential for CO_2 Storage in the Indian Subcontinent.* Report No.2008/2. Cheltenham, UK

[25] Jain, K. S., Agarwal, P. K. & Singh, V. P. (2007). *Hydrology and Water Resources of India.* Dordrecht, The Netherlands: Springer.

[26] Kumar, R., Singh. R. D. & Sharma K. D. (September 2005). "Water Resources of India." *Current Science. Volume, 89,* No 5

[27] Lardon, L., Helias, A., Sialve, B., Steyer, J. P. & Bernard, O. (2009). "Life-Cycle Assessment of Biodiesel Production from Microalgae." *Environmental Science and Technology, (43, 17)*; pp. 6475–6481.

[28] Meier, R. L. (1955). "Biological Cycles in the Transformation of Solar Energy into Useful Fuels." Daniels, F.; Duffie, J.A., eds. *Solar Energy Research,* Madison, WI: University of Wisconsin Press; pp. 179–1 83.

[29] Ministry of New and Renewable Energy (MNRE). (2009). "National Policy on Biofuels", New Delhi, India. http://www.mnre.gov.in/policy. Accessed January 13, 2010

[30] Ministry of Petroleum and Natural Gas (MPNG). (2009). *"Basic Statistics on Indian Petroleum and Natural Gas 2008-09."* New Delhi, India.

[31] Ministry of Rural Development (Department of Land Resources) and National Remote Sensing Agency (MRD-NRSA). (2005). *"Wastelands Atlas of India."* http://dolr.nic.in/WastelandsAtlas2005/Wasteland_Atlas_2005.pdf. Accessed May 27, 2010.

[32] Ministry of Water Resources (MWR). (2009). *"National Water Resources at a Glance."* http://wrmin.nic.in/index2.asp?sublinkid=290&langid=1&slid=412. Accessed November.

[33] Muneer, T., Asif, M. & Munawwar, S. (2004). "Sustainable Production of Solar Electricity with Particular Reference to the Indian Economy." *Renewable and Sustainable Energy Reviews*, Volume, *9*, Issue 5, October 2005, Pages 444-473

[34] U.S. National Aerospace and Space Administration (NASA). (2009). "NASA Satellites Unlock Secret to Northern India's Vanishing Water." http://www.nasa.gov/topics/earth/features/india water.html. Accessed October 2009.

[35] National Informatics Centre, National Portal of India. (2005). "Rivers." http://india.gov.in/knowindia/rivers.php. Accessed November 2009.

[36] National Institute of Hydrology (NIH). (2009). "Evaporation in India." http://www.nih.ernet.in/nih rbis/india information/evaporation.htm. Accessed November.

[37] New Agriculturist. (May 1999). "Country Profile – India." http://www.new-ag.info/country/profile.php?a=883 . Accessed May 27, 2010.

[38] The Oil Drum. (2009). "Produced Water, GOSPs and Saudi Arabia." http://www.theoildrum.com/node/6052 . Accessed May 27, 2010.

[39] Ravindranath, N. H., Somashekar, H. I., Nagaraja, M. S., Sudha, P., Sangeetha, G., Bhattacharya, S. C. & Salam, Abdul P. (2005). "Assessment of Sustainable Non-Plantation Biomass Resources Potential for Energy in India." *Biomass and Bioenergy, (29)*, pp. 178–190.

[40] Richmond, A. (2004). "Mass Cultivation of Microalgae." *Handbook of Microalgal Culture.* Blackwell Science Ltd.; pp. 95-252.

[41] Rubin, E. (2005). "Carbon Dioxide Capture and Storage." IPCC Technical Summary.

[42] Sheehan, J., Dunahay, T., Benemann, J. & Roessler, P. (1998). *A Look Back at the U.S. Department of Energy's Aquatic Species Program – Biodiesel from Algae.* NREL/TP-580- 24190. U.S. Department of Energy's Office of Fuels Development, Washington DC.

[43] Singh, N. et al. (2005). Indian Institute of Tropical Meteorology. "*Spatial Variation of Moisture Regions and Rainfall Zones of India During 1871-2003.*" Pune, Maharashtra, India.

[44] Prasad, S. Thenkabail, Venkateswarlu Dheeravath, Chandrashekhar M. Biradar, Obi Reddy P. Gangalakunta, Praveen Noojipady, Chandrakantha Gurappa, Manohar Velpuri, Muralikrishna Gummaand, Yuanjie Li (2009). "Irrigated Area Maps and Statistics of India Using Remote Sensing and National Statistics." *Remote Sensing Journal, (1)*, pp. 50-67.

[45] Ugwu, C. U., Aoyagi, H. & Uchiyama, H. (2007). "Photobioreactors for Mass Cultivation of Algae." *Bioresource Technology, (99)*, pp. 402 1-4028.

[46] United Nations: Millennium Development Goals Indicators. (2009). "Carbon Dioxide Emissions 2006." http://mdgs.un.org/unsd/mdg/SeriesDetail.aspx?srid=749&crid= . Accessed January 2010.

[47] U.S. Department of Energy Biomass Program (USDOE). (2009). "National Algal Biofuels Technology Roadmap." https ://e-center.doe.gov/iips/faopor.nsf/UNID/79E3ABCACC9AC14A852575CA00799D99/$file/AlgalB iofuels_Roadmap_7.pdf. Accessed May 27, 2010.

[48] U.S. Geological Survey (USGS). (2010). "Saline Water" http://ga.water. Accessed January 2010.

[49] Warner, D. L. (2001). *Technical and Economic Evaluations of the Protection of Saline Groundwater Under the Safe Drinking Water Act and the UIC Regulations.* Report submitted to the Ground Water Protection Council, http://www.gwpc.org.

[50] Weyer, K. M., Bush, D. R., Darzins, A. & Wilson, B. D. (2009). "Theoretical Maximum Algal Oil Production." *Bioenergy Research* DOI 10.1007/s12155-009-9046-x.

[51] Yale School of Forestry and Environmental Studies (YSFES). (September 2009). "India's CO_2 Emissions To At Least Triple in Next 20 Years." http://e360.yale.edu/content/digest.msp?id=2038. Accessed May 27, 2010.

REPORT DOCUMENTATION PAGE		Form Approved OMB No. 0704-0188
The public reporting burden for this collection of information is estimated to average 1 hour per response, including the time for reviewing instructions, searching existing data sources, gathering and maintaining the data needed, and completing and reviewing the collection of information. Send comments regarding this burden estimate or any other aspect of this collection of information, including suggestions for reducing the burden, to Department of Defense, Executive Services and Communications Directorate (0704-0188). Respondents should be aware that notwithstanding any other provision of law, no person shall be subject to any penalty for failing to comply with a collection of information if it does not display a currently valid OMB control number. **PLEASE DO NOT RETURN YOUR FORM TO THE ABOVE ORGANIZATION.**		

1. REPORT DATE *(DD-MM-YYYY)* September 2010	2. REPORT TYPE Technical Report	3. DATES COVERED *(From - To)*
4. TITLE AND SUBTITLE Resource Evaluation and Site Selection for Microalgae Production in India		**5a. CONTRACT NUMBER** DE-AC36-08-GO28308
		5b. GRANT NUMBER
		5c. PROGRAM ELEMENT NUMBER
6. AUTHOR(S) A. Milbrandt and E. Jarvis		**5d. PROJECT NUMBER** NREL/TP-6A2-48380
		5e. TASK NUMBER BB07.9610
		5f. WORK UNIT NUMBER
7. PERFORMING ORGANIZATION NAME(S) AND ADDRESS(ES) National Renewable Energy Laboratory 1617 Cole Blvd. Golden, CO 80401-3393		**8. PERFORMING ORGANIZATION REPORT NUMBER** NREL/TP-6A2-48380
9. SPONSORING/MONITORING AGENCY NAME(S) AND ADDRESS(ES)		**10. SPONSOR/MONITOR'S ACRONYM(S)** NREL
		11. SPONSORING/MONITORING AGENCY REPORT NUMBER
12. DISTRIBUTION AVAILABILITY STATEMENT National Technical Information Service U.S. Department of Commerce 5285 Port Royal Road Springfield, VA 22161		
13. SUPPLEMENTARY NOTES		
14. ABSTRACT *(Maximum 200 Words)* The study evaluates climate conditions, availability of CO2 and other nutrients, water resources, and land characteristics to identify areas in India suitable for algae production. The purpose is to provide an understanding of the resource potential in India for algae biofuels production and to assist policymakers, investors, and industry developers in their future strategic decisions.		
15. SUBJECT TERMS algae, biomass resources, India, biofuels, CO2		

16. SECURITY CLASSIFICATION OF:			**17. LIMITATION OF ABSTRACT**	**18. NUMBER OF PAGES**	**19a. NAME OF RESPONSIBLE PERSON**
a. REPORT Unclassified	**b. ABSTRACT** Unclassified	**c. THIS PAGE** Unclassified	UL		**19b. TELEPHONE NUMBER** *(Include area code)*

End Notes

[1] Green diesel is chemically different from biodiesel. It is composed of hydrocarbons with chemical properties identical to petroleum diesel, while biodiesel is not a pure hydrocarbon, but rather a mixture of fatty acid methyl esters (FAMEs). One method for producing green diesel is through hydroprocessing, the process of reacting the lipid feedstock with hydrogen under high temperature and pressure. Another method is through indirect liquefaction: first, the biomass is converted to a syngas, a gaseous mixture rich in hydrogen and carbon monoxide; then, the syngas is catalytically converted to liquids. The production of liquids is accomplished using Fischer-Tropsch (FT) synthesis.

[2] 1 tonne = 7.4 barrels, assuming an average density of 0.85 g/cm^3 (or kg/L or t/m^3) and 159 L/bbl (oil barrel; 1 non-beer fluid barrel = 119 L)

[3] India's HSD (high speed diesel) consumption in 2008–2009 was 52 Mt (MPNG 2009).

[4] Country area is about 328 Mha.

[5] Slope is steepness of the landscape. It is measured in degrees or as a percentage. Percent Slope = Rise/Run* 100. For example, a rise of 2 meters over a distance of 100 meters describes a 2% slope. In general, 0%–5% slope is considered flat area, gentle slope is 6%–1 0%, moderate slope is 11 %–25%, and steep slope is 25% and greater.

[6] Assuming algal oil productivity between 4–10 t/ha/yr.

[7] India's HSD (high speed diesel) consumption in 2008–2009 was 52 Mt (MPNG 2009).

[8] Assuming for every 40 gallons of algal oil (about 0.13 tonnes), a tonne of CO_2 is consumed during the algal biomass production at 30% lipid algal content by weight (USDOE 2009).

[9] Stationary CO_2 sources emitted 638 Mt of CO_2 in 2007, nearly half of the total CO_2 emissions in India (CARMA 2007).

In: Exploring Renewable and Alternative Energy Use in India   ISBN: 978-1-61209-680-3
Editor: Jonathan R. Mulder   © 2012 Nova Science Publishers, Inc.

Chapter 3

NATIONAL POLICY ON BIOFUELS

Government of India

1.0. PREAMBLE

1.1 India is one of the fastest growing economies in the world. The Development Objectives focus on economic growth, equity and human well being. Energy is a critical input for socio-economic development. The energy strategy of a country aims at efficiency and security and to provide access which being environment friendly and achievement of an optimum mix of primary resources for energy generation. Fossil fuels will continue to play a dominant role in the energy scenario in our country in the next few decades. However, conventional or fossil fuel resources are limited, non-renewable, polluting and, therefore, need to be used prudently. On the other hand, renewable energy resources are indigenous, non-polluting and virtually inexhaustible. India is endowed with abundant renewable energy resources. Therefore, their use should be encouraged in every possible way.

1.2 The crude oil price has been fluctuating in the world market and has increased significantly in the recent past, reaching a level of more than $ 140 per barrel. Such unforeseen escalation of crude oil prices is severely straining various economies the world over, particularly those of the developing countries. Petro-based oil meets about 95% of the requirement for transportation fuels, and the demand has been steadily rising. Provisional estimates have indicated crude oil consumption in 2007-08 at about 156 million tonnes. The domestic crude oil is able to meet only about 23% of the demand, while the rest is met from imported crude.

1.3 India's energy security would remain vulnerable until alternative fuels to substitute/supplement petro-based fuels are developed based on indigenously produced renewable feedstocks. In biofuels, the country has a ray of hope in providing energy security. Biofuels are environment friendly fuels and their utilization would address global concerns about containment of carbon emissions. The transportation sector has been identified as a major polluting sector. Use of biofuels have, therefore, become compelling in view of the tightening automotive vehicle emission standards to curb air pollution.

1.4 Biofuels are derived from renewable bio-mass resources and, therefore, provide a strategic advantage to promote sustainable development and to supplement conventional energy sources in meeting the rapidly increasing requirements for transportation fuels associated with high economic growth, as well as in meeting the energy needs of India's vast rural population. Biofuels can increasingly satisfy these energy needs in an environmentally benign and cost-effective manner while reducing dependence on import of fossil fuels and thereby providing a higher degree of National Energy Security.

1.5 The growth of biofuels around the globe is spurred largely by energy security and environmental concerns and a wide range of market mechanisms, incentives and subsidies have been put in place to facilitate their growth. Developing countries, apart from these considerations, also view biofuels as a potential means to stimulate rural development and create employment opportunities. The Indian approach to biofuels, in particular, is somewhat different to the current international approaches which could lead to conflict with food security. It is based solely on non-food feedstocks to be raised on degraded or wastelands that are not suited to agriculture, thus avoiding a possible conflict of fuel vs. food security.

1.6 In the context of the International perspectives and National imperatives, it is the endeavour of this Policy to facilitate and bring about optimal development and utilization of indigenous biomass feedstocks for production of biofuels. The Policy also envisages development of the next generation of more efficient biofuel conversion technologies based on new feedstocks. The Policy sets out the Vision, medium term Goals, strategy and approach to biofuel development, and proposes a framework of technological, financial and institutional interventions and enabling mechanisms.

2.0. THE VISION AND GOALS

2.1 The Policy aims at mainstreaming of biofuels and, therefore, envisions a central role for it in the energy and transportation sectors of the country in coming decades. The Policy will bring about accelerated development and promotion of the cultivation, production and use of biofuels to increasingly substitute petrol and diesel for transport and be used in stationary and other applications, while contributing to energy security, climate change mitigation, apart from creating new employment opportunities and leading to environmentally sustainable development.

2.2 The Goal of the Policy is to ensure that a minimum level of biofuels become readily available in the market to meet the demand at any given time. An indicative target of 20% blending of biofuels, both for bio-diesel and bio-ethanol, by 2017 is proposed. Blending levels prescribed in regard to bio-diesel are intended to be recommendatory in the near term. The blending level of bio-ethanol has already been made mandatory, effective from October, 2008, and will continue to be mandatory leading upto the indicative target.

3.0. DEFINITIONS AND SCOPE

3.1 The following definitions of biofuels shall apply for the purpose of this Policy:
  i. 'biofuels' are liquid or gaseous fuels produced from biomass resources and used in place of, or in addition to, diesel, petrol or other fossil fuels for transport, stationary, portable and other applications;
  ii. 'biomass' resources are the biodegradable fraction of products, wastes and residues from agriculture, forestry and related industries as well as the biodegradable fraction of industrial and municipal wastes.
3.2 The scope of the Policy encompasses bio-ethanol, bio-diesel and other biofuels, as listed below:-
  i. 'bio-ethanol': ethanol produced from biomass such as sugar containing materials, like sugar cane, sugar beet, sweet sorghum, etc.; starch containing materials such as corn, cassava, algae etc.; and, cellulosic materials such as bagasse, wood waste, agricultural and forestry residues etc. ;
  ii. 'biodiesel': a methyl or ethyl ester of fatty acids produced from vegetable oils, both edible and non-edible, or animal fat of diesel quality; and ,
  iii. other biofuels: biomethanol, biosynthetic fuels etc.

4.0. STRATEGY AND APPROACH

4.1 The focus for development of biofuels in India will be to utilize waste and degraded forest and non-forest lands only for cultivation of shrubs and trees bearing non-edible oil seeds for production of bio-diesel. In India, bio-ethanol is produced mainly from molasses, a by-product of the sugar industry. In future too, it would be ensured that the next generation of technologies is based on non-food feedstocks. Therefore, the issue of fuel vs. food security is not relevant in the Indian context.
4.2 Cultivators, farmers, landless labourers etc. will be encouraged to undertake plantations that provide the feedstock for bio-diesel and bio-ethanol. Corporates will also be enabled to undertake plantations through contract farming by involving farmers, cooperatives and Self Help Groups etc. in consultation with Panchayats, where necessary. Such cultivation / plantation will be supported through a Minimum Support Price for the non-edible oil seeds used to produce bio-diesel.
4.3 In view of the current direct and indirect subsidies to fossil fuels and distortions in energy pricing, a level playing field is necessary for accelerated development and utilization of biofuels to subserve the Policy objectives. Appropriate financial and fiscal measures will be considered from time to time to support the development and promotion of biofuels and their utilization in different sectors.
4.4 Research, development and demonstration will be supported to cover all aspects from feedstock production and biofuels processing for various end-use applications. Thrust will also be given to development of second generation biofuels and other new feedstocks for production of bio-diesel and bio-ethanol.

5.0. Interventions and Enabling Mechanisms

Plantations

5.1 Plantations of trees bearing non-edible oilseeds will be taken up on Government/ community wasteland, degraded or fallow land in forest and non-forest areas. Contract farming on private wasteland could also be taken up through the Minimum Support Price mechanism proposed in the Policy. Plantations on agricultural lands will be discouraged.

5.2 There are over 400 species of trees bearing non-edible oilseeds in the country. The potential of all these species will be exploited, depending on their techno-economic viability for production of biofuels. Quality seedlings would be raised in the nurseries of certified institutions / organizations identified by the States for distribution to the growers and cultivators.

5.3 In all cases pertaining to land use for the plantations, consultations would be undertaken with the local communities through Gram Panchayats/ Gram Sabhas, and with Intermediate Panchayats and District Panchayat where plantations of non-edible oil seed bearing trees and shrubs are spread over more than one village or more than one block/ taluk. Further, the provisions of PESA would be respected in the Fifth Schedule Areas.

5.4 A major instrument of this Policy is that a Minimum Support Price (MSP) for oilseeds should be announced and implemented with a provision for its periodic revision so as to ensure a fair price to the farmers. The details about implementation of the MSP mechanism will be worked out carefully after due consultations with concerned Government agencies, States and other stakeholders. It will then be considered by the Biofuel Steering Committee and decided by the National Biofuels Co-ordination Committee proposed to be set up under this Policy. The Statutory Minimum Price (SMP) mechanism prevalent for sugarcane procurement will also be examined for extending such a mechanism for oilseeds to be utilized for production of bio-diesel by the processing units. Payment of SMP would be the responsibility of the bio-diesel processors. Different levels of Minimum Support Price for oilseeds has already been declared by certain States.

5.5 Employment provided in plantations of trees and shrub bearing nonedible oilseeds will be made eligible for coverage under the National Rural Employment Guarantee Programme (NREGP).

Processing

5.6 Ethanol is mainly being produced in the country at present from molasses, which is a by-product of the sugar industry. 5% blending of ethanol with gasoline has already been taken up by the Oil Marketing Companies (OMCs) in 20 States and 4 Union Territories. 10% mandatory blending of ethanol with gasoline is to become effective from October, 2008 in these States. In order to augment availability of ethanol and reduce over supply of sugar, the sugar industry has been permitted to produce ethanol directly from sugarcane juice. The sugar and distillery industry will be further encouraged to augment production of ethanol to meet the blending requirements prescribed from time to time, while ensuring that this does not in any way create supply constraints in production of sugar or availability of ethanol for industrial use.

5.7 Setting up of processing units by industry for bio-oil expelling/extraction and transesterification for production of bio-diesel will be encouraged. While it is difficult to exactly specify the percentage of bio-diesel to be blended with diesel in view of the uncertainty in the availability of bio-diesel at least in the initial stages, blending will be permitted upto certain prescribed levels, to be recommendatory initially and made mandatory in due course. Gram/Intermediate Panchayats would also be encouraged to create facilities at the village level for extraction of bio-oil, which could then be sold to bio-diesel processing units.

5.8 The prescribed blending levels will be reviewed and moderated periodically as per the availability of bio-diesel and bio-ethanol. A National Registry of feedstock availability, processing facilities and offtake will be developed and maintained to provide necessary data for such reviews with a view to avoid mismatch between supply and demand.

5.9 In order to take care of fluctuations in the availability of biofuels, OMCs will be permitted to bank the surplus quantities left after blending of bio-diesel and bio-ethanol in a particular year, and to carry it forward to the subsequent year when there may be a shortfall in their availability to meet the prescribed levels.

5.10 The blending would have to follow a protocol and certification process, and conform to BIS specification and standards, for which the processing industry and OMCs would need to jointly set up an appropriate mechanism and the required facilities. Section 52 of the Motor Vehicles Act already allows conversion of an existing engine of a vehicle to use biofuels. Engine manufacturers would need to suitably modify the engines to ensure compatibility with biofuels, wherever necessary.

Distribution & Marketing of Biofuels

5.11 The responsibility of storage, distribution and marketing of biofuels would rest with OMCs. This shall be carried out through their existing storage and distribution infrastructure and marketing networks, which may be suitably modified or upgraded to meet the requirements for biofuels.

5.12 In the determination of bio-diesel purchase price, the entire value chain comprising production of oil seeds, extraction of bio-oil, its processing, blending, distribution and marketing will have to be taken into account. The Minimum Purchase Price (MPP) for bio-diesel by the OMCs will be linked to the prevailing retail diesel price. The MPP for bio-ethanol, will be based on the actual cost of production and import price of bio-ethanol. The MPP, both for bio-diesel and bio-ethanol will be determined by the Biofuel Steering Committee and decided by the National Biofuel Coordination Committee. In the event of diesel or petrol price falling below the MPP for bio-diesel and bio-ethanol, OMCs will be duly compensated by the Government.

Financing

5.13 Plantation of non-edible oil bearing plants, the setting up of oil expelling/extraction and processing units for production of bio-diesel and creation of any new infrastructure for storage and distribution would be declared as a priority sector for the purposes of lending by financial institutions and banks. National Bank of Agriculture and Rural Development (NABARD) would provide re-financing towards loans to farmers for plantations. Indian Renewable Energy Development Agency (IREDA), Small Industries Development Bank of India (SIDBI) and other financing agencies as well as commercial banks would be actively involved in providing finance for various activities under the entire biofuel value chain, at different levels.

5.14 Multi-lateral and bi-lateral funding would be sourced, where possible for biofuel development. Carbon financing opportunities would also be explored on account of avoidance of CO_2 emissions through plantations and use of biofuels for various applications.

5.15 Investments and joint ventures in the biofuel sector are proposed to be encouraged. Biofuel technologies and projects would be allowed 100% foreign equity through automatic approval route to attract Foreign Direct Investment (FDI), provided biofuel is for domestic use only, and not for export. Plantations would not be open for FDI participation.

Financial and Fiscal Incentives

5.16 Financial incentives, including subsidies and grants, may be considered upon merit for new and second generation feedstocks; advanced technologies and conversion processes; and, production units based on new and second generation feedstocks. If it becomes necessary, a National Biofuel Fund could be considered for providing such financial incentives.

5.17 As biofuels are derived from renewable biomass resources they will be eligible for various fiscal incentives and concessions available to the New and Renewable Energy Sector from the Central and State Governments.

5.18 Bio-ethanol already enjoys concessional excise duty of 16% and bio-diesel is exempted from excise duty. No other Central taxes and duties are proposed to be levied on bio-diesel and bio-ethanol. Custom and excise duty concessions would be provided on plant and machinery for production of bio-diesel or bio-ethanol, as well as for engines run on biofuels for transport, stationary and other applications, if these are not manufactured indigenously.

Research & Development and Demonstration

5.19 A major thrust would be given through this Policy to Innovation, Research & Development and Demonstration in the field of biofuels. Research and Development will focus on plantations, biofuel processing and production technologies, as well as on maximizing efficiencies of different end-use applications and utilization of by-products. High priority will be accorded to indigenous R&D and technology development based on local feedstocks and needs, which would be benchmarked with international efforts and patents would be registered, wherever possible. Multi-institutional, time-bound research programmes with clearly defined goals and milestones would be developed and supported.

5.20 Intensive R&D work would be undertaken in the following areas:

(a): Biofuel feed-stock production based on sustainable biomass with active involvement of local communities through non-edible oilseed bearing plantations on wastelands to include *inter-alia* production and development of quality planting materials and high sugar containing varieties of sugarcane, sweet sorghum, sugar beet, cassava, etc.

(b): Advanced conversion technologies for first generation biofuels and emerging technologies for second generation biofuels including conversion of ligno-cellulosic materials to ethanol such as crop residues, forest wastes and algae, biomass-to-liquid (BTL) fuels, bio-refineries, etc.

(c): Technologies for end-use applications, including modification and development of engines for the transportation sector based on a large scale centralized approach, and for stationary applications for motive power and electricity production based on a decentralized approach.

(d): Utilisation of by-products of bio-diesel and bio-ethanol production processes such as oil cake, glycerin, bagasse, etc.

5.21 Demonstration Projects will be set up for biofuels, both for bio-diesel and bio-ethanol production, conversion and applications based on state-of-the-art technologies through Public Private Partnership (PPP).

5.22 For R&D and demonstration projects, grants would be provided to academic institutions, research organizations, specialized centers and industry. Strengthening of existing R&D centers and setting up of specialized centers in high technology areas will also be considered. Linkages would be established between the organizations / agencies undertaking technology development and the user organizations. Transfer of know-how would be facilitated to industry. Participation by industry in R&D and technology development will be encouraged with increased investment by industry with a view to achieve global competitiveness.

5.23 In regard to Research and Development in the area of biofuels, a Subcommittee under the Biofuel Steering Committee proposed in this Policy comprising Department of Bio-Technology, Ministry of Agriculture, Ministry of New and Renewable Energy and Ministry of Rural Development would be constituted, led by Department of Bio-Technology and coordinated by the Ministry of New and Renewable Energy.

6.0. QUALITY STANDARDS

6.1 Development of test methods, procedures and protocols would be taken up on priority alongwith introduction of standards and certification for different biofuels and end use applications. The Bureau of Indian Standards (BIS) has already evolved a standard (IS-15607) for Bio-diesel (B 100), which is the Indian adaptation of the American Standard ASTM D-6751 and European Standard EN-14214. BIS has also published IS: 2796: 2008 which covers specification for motor gasoline blended with 5% ethanol and motor gasoline blended with 10% ethanol.

6.2 The Bureau of Indian Standards (BIS)would review and update the existing standards, as well as develop new standards in a time-bound manner for devices and systems for various end-use applications for which standards have not yet been prepared, at par with international standards. Guidelines for product performance and reliability would also be developed and institutionalized in consultation with all relevant stakeholders. Standards would be strictly enforced and proper checks would be carried out by a designated agency on the quality of the biofuel being supplied.

7.0. International Cooperation

7.1 International scientific and technical cooperation in the area of biofuel production, conversion and utilization will be established in accordance with national priorities and socio-economic development strategies and goals. Modalities of such cooperation may include joint research and technology development, field studies, pilot scale plants and demonstration projects with active involvement of research institutions and industry on either side. Technology induction/ transfer would be facilitated, where necessary, with time-bound goals for indigenisation and local manufacturing. Appropriate bilateral and multi-lateral cooperation programmes for sharing of technologies and funding would be developed, and participation in international partnerships, where necessary, will also be explored.

8.0. Import and Export of Biofuels

8.1 Import of biofuels would only be permitted to the extent necessary, and will be decided by the National Biofuel Coordination Committee proposed under this Policy. Duties and taxes would be levied on the imports so as to ensure that indigenously produced biofuels are not costlier than the imported biofuels. Import of Free Fatty Acid (FFA) oils will not be permitted for production of biofuels.

8.2 Export of biofuels would only be permitted after meeting the domestic requirements and would be decided by the National Biofuel Coordination Committee.

9.0. Role of States

9.1 The role and active participation of the States is crucial in the planning and implementation of biofuel programmes. The State Governments would be asked to designate an existing agency, or create a new agency suitably empowered and funded to act as nodal agency for development and promotion of biofuels in their States. Certain States have already set up such agencies. Other concerned agencies, panchayati raj institutions, forestry departments, universities, research institutions etc. would also need to be associated in these efforts. While a few States have announced policies for biofuel development, other States would also need to announce suitable policies in a time-bound manner in line with the broad contours and provisions of this National Policy.

9.2 State Governments would also be required to decide on land use for plantation of non-edible oilseed bearing plants or other feedstocks of biofuels, and on allotment of Government wasteland, degraded land for raising such plantations. Creation of necessary infrastructure would also have to be facilitated to support biofuel projects across the entire value chain.

10.0. AWARENESS AND CAPACITY BUILDING

10.1 Support will be provided for creation of awareness about the role and importance of biofuels in the domestic energy sector, as well as for wide dissemination of information about its potential and opportunities in upgrading the transportation infrastructure and supporting the rural economy.

10.2 Significant thrust would be provided to capacity building and training and development of human resources. Universities, Polytechnics and Industrial Training Institutes will be encouraged to introduce suitable curricula to cater to the demand for trained manpower at all levels in different segments of the biofuel sector. Efforts will also be directed at enhancing and expanding consultancy capabilities to meet the diverse requirements of this sector.

11.0. INSTITUTIONAL MECHANISMS

11.1 Under the Allocation of Business Rules, the Ministry of New & Renewable Energy has been given the responsibility of Policy and overall Coordination concerning biofuels. Apart from this, the Ministry has also been given the responsibility to undertake R&D on various applications of biofuels. Responsibilities have also been allocated to other Ministries viz. Ministry of Environment & Forests, Ministry of Petroleum & Natural Gas, Ministry of Rural Development and Ministry of Science & Technology to deal with different aspects of biofuel development and promotion in the country.

11.2 In view of a multiplicity of departments and agencies, it is imperative to provide High-level co-ordination and policy guidance / review on different aspects of biofuel development, promotion and utilization. For this purpose, it is proposed to set up a National Biofuel Coordination Committee (NBCC) headed by the Prime Minister. Ministers from concerned Ministries would be Members of this Committee. The Committee would meet periodically to provide overall coordination, effective end-to-end implementation and monitoring of biofuel programmes.

11.3 The National Biofuel Coordination Committee will have the following composition:

Chairman: Prime Minister of India
Members:

i.   Deputy Chairman, Planning Commission
ii. Minister of New and Renewable Energy
iii. Minister of Rural Development
iv. Minister of Agriculture
v.   Minister of Environment & Forests
vi. Minister of Petroleum & Natural Gas
vii. Minister of Science & Technology
viii. Secretary, Ministry of New and Renewable Energy -**Convener**

Coordinating Ministry: Ministry of New and Renewable Energy
11.4 In order to provide effective guidance and to oversee implementation of the Policy on a regular and continuing basis, it is proposed to set up a Biofuel Steering Committee headed by the Cabinet Secretary, and comprising Secretaries of concerned departments.
11.5 The Biofuel Steering Committee will have the following composition:-

Chairman: Cabinet Secretary
Members:
i.   Secretary, Ministry of Finance
ii. Secretary, Ministry of Rural Development, Department of Land Resources
iii. Secretary, Department of Agricultural Research and Education
iv. Secretary, Ministry of Environment & Forests
v.   Secretary, Ministry of Petroleum & Natural Gas
vi. Secretary, Department of Science & Technology
vii. Secretary, Ministry of Panchayati Raj
viii. Secretary, Department of Biotechnology
ix. Secretary, Planning Commission
x.   Secretary, Department of Scientific & Industrial Research
xi. Secretary, Ministry of New & Renewable Energy **Member Secretary**

Coordinating Ministry: Ministry of New and Renewable Energy
11.6 In order to enable the Ministry of New & Renewable Energy to effectively carry out its role as the coordinating Ministry for the National Biofuel Progamme, it will be necessary for it to be suitably strengthened through augmentation of its manpower with the flexibility of hiring external professional manpower and services.

CHAPTER SOURCES

The following chapters have been previously published:

Chapter 1 – This is an edited reformatted and augmented version of a National Renewable Energy Laboratory publication, report NREL/TP-6A20-48948, dated October 2010.

Chapter 2 – This is an edited reformatted and augmented version of a National renewable Energy Laboratory publication, report NREL/TP-6A2-48380, dated September 2010.

Chapter 3 – This is an edited reformatted and augmented version of a Government of India, Ministry of New and Renewable Energy publication, report *National Policy on Bio Fuels*.

INDEX

A

access, vii, 4, 6, 8, 9, 11, 12, 18, 25, 27, 29, 31, 75, 78, 79, 81, 83, 87, 94, 96, 117, 146, 205
accounting, 10, 11
acid, 130, 203
adaptability, 86
adaptation, 6, 13, 25, 84, 86, 87, 212
advancement, 11
advancements, 7
aerosols, 132
Africa, 52, 104
agencies, 6, 28, 78, 87, 92, 118, 208, 210, 211, 213
agricultural sector, 19, 146
agriculture, 19, 78, 90, 126, 148, 206, 207
algae, 74, 124, 125, 126, 127, 128, 129, 130, 131, 132, 134, 135, 142, 144, 145, 146, 148, 149, 150, 151, 152, 155, 156, 157, 158, 159, 160, 161, 207, 211
anaerobic digesters, 64
annual rate, 18, 36
Appropriate Rural Technology Institute, 1, 78
appropriate technology, 87
aquaculture, 135, 151
Argentina, 102
Asia, 1, 34, 39, 86, 90, 103, 108, 110, 119, 134
Asia Pacific Centre for the Transfer of Technology, 1, 90
Asian countries, 101
Asian Development Bank (ADB), 28
assessment, 7, 34, 35, 41, 55, 66, 78, 153, 199
assets, 94, 106, 111
athletes, 27
atmosphere, 97, 125, 128, 132, 148, 150, 157, 159, 161
authority, 28

avoidance, 210
awareness, 53, 56, 89, 91, 213

B

balance sheet, 91, 94
Bangladesh, 96, 102
banking, 28, 59, 79, 97
banks, 89, 94, 97, 103, 120, 210
barriers, 55
base, 10, 17, 49, 83, 125, 157, 161
batteries, 96
BEE, 1, 89, 119
beer, 203
Beijing, 100, 101
benefits, 4, 7, 14, 24, 29, 37, 38, 49, 69, 81, 83, 94, 130, 131, 142, 148
benign, 57, 130, 206
Bharat Heavy Electricals Ltd, 1, 40
biochemistry, 127
biodiesel, 9, 10, 11, 27, 63, 71, 72, 73, 74, 83, 92, 126, 128, 130, 157, 203, 207
biodiversity, 127, 152
bioenergy, 8, 9, 10, 24, 28, 70, 95
biofuel, 11, 63, 71, 73, 92, 206, 210, 212, 213
biogas, 8, 9, 25, 63, 64, 65, 66, 67, 70, 76, 86, 93, 94, 116, 131
Biogas Distributed/Grid Power Generation Program, 1, 65
biomass, 4, 6, 8, 9, 17, 20, 25, 27, 28, 52, 55, 62, 63, 67, 68, 69, 70, 71, 75, 77, 78, 79, 84, 86, 87, 96, 98, 125, 126, 128, 129, 130, 131, 142, 150, 156, 157, 161, 203, 206, 207, 210, 211
biopower, 11, 69
biotechnology, 90
blends, 92

boilers, 69
borrowers, 94
Brazil, 54, 74, 102, 136
breakdown, 21
building code, 8, 56
business model, 4, 9, 24, 25, 66, 78, 79, 80, 86, 90
businesses, 9, 17, 79, 93, 96
buyers, 18, 103
by-products, 211

C

Cabinet, 214
CAD, 102
campaigns, 89
capacity building, 9, 24, 53, 56, 62, 83, 89, 90, 94, 100, 213
carbohydrates, 131
carbon, 5, 10, 13, 74, 83, 91, 96, 98, 103, 106, 123, 125, 127, 128, 148, 159, 161, 203, 205
carbon dioxide, 13, 123, 127
carbon emissions, 5, 148, 205
carbon monoxide, 203
carotenoids, 131
case study, 67, 116
cash, 93, 94, 97
cash flow, 93, 94
category a, 58, 59
cattle, 9, 55, 63, 64
cell division, 127
Census, 17, 106
Central Electricity Authority, 2, 4, 14
Central Electricity Regulatory Commission, 2, 6, 28
Central Finance Assistance, 2
Centre for Wind Energy Technology, 2, 6, 28
certificate, 101
certification, 13, 89, 209, 211
certified emission reduction credit, 2, 13
challenges, vii, 4, 5, 11, 17, 22, 23, 26, 56, 99, 124, 126, 131, 144, 151, 155
chemical, 81, 129, 203
chemical properties, 203
chemicals, 81, 148
Chile, 102
China, 10, 14, 16, 34, 37, 48, 53, 54, 62, 90, 91, 98, 101, 102, 103, 109, 115, 120, 200
CIA, 136, 199
cities, 24, 74, 90, 101, 148, 149
citizens, 12
City, 104

civil society, 100, 101
Clean Development Mechanism, 2, 6, 12, 13, 97, 106
clean energy, 10, 13, 27, 87, 88, 89, 90, 91, 92, 94, 95, 96, 99, 102
clean technology, 94, 98
cleaning, 55
clients, 96
climate, 5, 10, 11, 12, 13, 24, 29, 83, 84, 90, 97, 104, 119, 125, 132, 134, 135, 136, 141, 146, 157, 160, 206
closure, 44, 51
clusters, 9, 64, 84, 90, 149
CO2, 13, 67, 123, 125, 127, 128, 129, 130, 131, 132, 135, 148, 150, 151, 157, 159, 160, 161, 199, 200, 202, 203, 210
coal, 11, 13, 17, 19, 23, 25, 40, 44, 52, 75, 82, 88, 148, 151
coastal region, 135
cogeneration, 9, 10, 25, 29, 30, 31, 63, 67, 68, 69, 70, 81, 82, 86, 93, 97
collaboration, 1, 47, 75, 89, 90
collateral, 94
colleges, 88
combustion, 9, 64, 67, 70, 86, 97, 148
commercial, 4, 8, 10, 11, 13, 14, 16, 17, 19, 27, 34, 37, 51, 53, 55, 62, 63, 71, 72, 75, 77, 78, 87, 89, 91, 92, 94, 97, 98, 103, 148, 156, 210
commercial bank, 91, 94, 103, 210
communication, 115
communities, 18, 30, 65, 70, 74, 80, 208, 211
community, 10, 18, 57, 64, 65, 78, 80, 85, 95, 208
compact fluorescent lamp, 2
compatibility, 209
compensation, 84, 102, 155
competition, 5, 12, 25, 28, 29, 39, 40, 66, 85, 153, 159
competition policy, 28
competitiveness, 10, 211
complement, 87, 148
complexity, 129
compliance, 29
composition, 21, 130, 214
concentrated solar power, 2, 25
conditioning, 26, 121
conductor, 46
conference, 11, 100
conflict, 206
Congress, 119
connectivity, 31
conservation, 56, 89, 93, 139

constituents, 146

construction, 17, 21, 40, 58, 60, 61, 94, 104, 129, 148, 151, 157

consumers, 4, 18, 23, 37, 87, 88, 96

consumption, 9, 11, 14, 15, 16, 18, 19, 21, 23, 24, 26, 31, 52, 71, 107, 125, 126, 129, 142, 144, 155, 157, 203, 205

contamination, 129

convention, 97

cooking, 8, 17, 27, 55, 62, 63, 64, 65, 68, 72, 75, 76, 106, 126

cooling, 26, 63, 104, 110, 129

cooperation, 14, 23, 86, 89, 212

coordination, 28, 62, 73, 100, 213

corporate social responsibility, 2

cost, 4, 10, 11, 12, 13, 17, 24, 25, 28, 29, 33, 39, 41, 43, 44, 45, 47, 48, 53, 55, 59, 61, 73, 82, 84, 85, 87, 89, 92, 93, 97, 102, 103, 125, 127, 128, 129, 130, 131, 135, 142, 146, 157, 161, 206, 209

cotton, 68, 148

credit market, 96, 97

criticism, 13

crop, 17, 63, 68, 70, 71, 148, 151, 157, 211

crop production, 151

crop residue, 17, 63, 211

crops, vii, 83, 124, 126, 127, 128, 151

crude oil, vii, 124, 126, 146, 205

crystalline, 45, 46, 47, 103

cultivation, 9, 72, 73, 127, 129, 131, 132, 134, 145, 151, 157, 159, 206, 207

culture, 127, 128, 129, 131, 156

curricula, 213

customers, 4, 17, 75

cyclones, 135

D

Danish International Development Agency, 2, 95

data collection, 61

database, 117

Decentralized Distributed Generation, 2

deduction, 88, 107

deficit, 4, 11, 17, 18, 23, 53, 135

deforestation, 139

degradation, 139

Delhi International Renewable Energy Conference, 2, 11

Department of Agriculture, 116, 117

Department of Energy, 1, 121, 123, 127, 201

Department of Non-conventional Energy Sources, 2

depreciation, 24, 29, 37, 38, 45, 52, 69, 94

depth, 41, 142, 156

destruction, 57

developed countries, 24, 25, 86, 97

developed nations, 5, 12

developing countries, 13, 104, 205

developing nations, 86

development banks, 103

diatoms, 127, 151

diesel engines, 126

diesel fuel, 26, 74, 150

diffusion, 84

digestion, 64, 131

direct normal irradiance, 2, 49

disbursement, 98

displacement, vii, 57, 124, 126

distortions, 207

distributed applications, 75

distribution, 6, 7, 18, 23, 25, 28, 29, 30, 31, 32, 33, 35, 47, 66, 67, 76, 77, 84, 86, 93, 96, 143, 146, 151, 153, 155, 208, 209, 210

DOI, 202

donors, 89

draft, 1, 7, 32

drainage, 125, 136, 146, 148, 158, 160

drinking water, 136

drying, 128

dust storms, 132, 135

E

earnest money deposit, 2, 44

economic crisis, 91

economic development, vii, 11, 76, 140, 148, 205, 212

economic growth, vii, 4, 12, 14, 19, 148, 205, 206

economics, 87, 127, 129, 130, 131, 142

economies of scale, 5, 12, 46, 51, 59, 103

education, 10, 84

effluent, 64, 125, 146, 151, 157, 161

electricity, 4, 5, 6, 7, 8, 9, 10, 11, 12, 13, 14, 17, 18, 19, 21, 23, 24, 25, 26, 27, 28, 29, 30, 31, 33, 37, 38, 41, 45, 47, 49, 51, 52, 53, 55, 62, 63, 64, 65, 66, 67, 68, 69, 70, 74, 75, 76, 77, 78, 79, 80, 81, 82, 84, 85, 89, 97, 101, 103, 104, 108, 117, 118, 128, 136, 211

emergency, 81

emission, 2, 13, 97, 99, 206

employees, 67

employment, 65, 206

employment opportunities, 206

energy, vii, 4, 5, 6, 7, 8, 9, 10, 11, 12, 13, 14, 15, 16, 17, 18, 19, 20, 23, 24, 25, 26, 27, 28, 29, 30, 31, 32, 33, 34, 38, 41, 43, 44, 49, 52, 53, 56, 60, 61, 62, 66, 68, 69, 70, 75, 76, 77, 78, 79, 81, 84, 85, 86, 88, 89, 90, 91, 92, 93, 94, 95, 96, 97, 98,99, 100, 101, 102, 104, 106, 109, 110, 113, 115, 119, 120, 124, 126, 127, 128, 129, 130, 135, 144, 148, 205, 206, 207, 213

energy consumption, 16, 19, 90, 148

energy efficiency, 6, 14, 28, 90, 96, 98

energy prices, 119

energy supply, vii, 4, 8, 11, 14, 17, 27, 41, 75

enforcement, 37, 38, 81

engineering, 49, 62, 81, 125, 159, 161

engineer-procure-construct, 2, 44

entrepreneurs, 65, 78, 79, 91, 93

environment, vii, 2, 6, 10, 11, 41, 45, 47, 51, 83, 85, 86, 91, 96, 127, 128, 132, 135, 148, 205

environmental conditions, 40

environmental impact, 131, 135

EPC, 2, 44, 46, 51, 70, 85

equipment, 7, 23, 38, 39, 40, 44, 53, 61, 62, 69, 88, 94, 95, 110

equity, vii, 87, 89, 91, 94, 95, 98, 102, 205, 210

ester, 207

ethanol, 2, 9, 11, 27, 71, 72, 73, 74, 92, 116, 126, 131, 206, 207, 208, 209, 210, 211, 212

ethanol blended petrol, 2, 73

Europe, 39, 52, 104, 113

European Commission, 106

European Investment Bank, 103

European Union (EU), 2, 53, 54, 89, 101, 102, 103

evacuation, 51

evaporation, 129, 131, 132, 135, 138, 145, 146, 201

execution, 43, 45, 58, 61, 84

expenditures, 111

Expert Group on Technology Transfer, 2, 90

expertise, 86

export market, 8, 46

exposure, 103

externalities, 127, 130

extraction, 125, 128, 129, 130, 131, 142, 146, 160, 209, 210

F

---

factories, 17, 65, 82

families, 96

farmers, 19, 23, 70, 207, 208, 210

farms, 70, 125, 135, 146, 148, 151, 155, 157, 158, 159, 160, 161

fat, 27, 63, 72, 126, 207

fatty acids, 128, 131, 207

FDI, 210

federal government, 10, 12

feedstock, 9, 27, 63, 64, 67, 71, 72, 74, 86, 130, 157, 203, 207, 209

fermentation, 9, 27, 71

fertilizers, 66, 81

filtration, 129

financial, 6, 9, 10, 12, 13, 14, 17, 23, 24, 26, 27, 28, 40, 42, 44, 45, 46, 51, 56, 58, 59, 61, 62, 63, 65, 66, 69, 73, 79, 84, 87, 89, 90, 91, 92, 93, 94, 95, 96, 98, 99, 102, 123, 206, 207, 210

financial crisis, 102

financial incentives, 9, 27, 42, 46, 61, 63, 69, 73, 89, 210

financial institutions, 95, 98, 99, 210

financial markets, 100

financial resources, 14

financial support, 6, 10, 26, 28, 65, 66, 84, 90, 94, 123

first generation, 211

flex, 74

flexibility, 130, 214

flocculation, 129

floods, 132, 135

flotation, 129

fluctuations, 17, 141, 209

flue gas, 128, 146, 157

fluid, 203

fluidized bed, 69

food, 63, 64, 65, 71, 127, 131, 148, 151, 153, 206, 207

force, 76

forecasting, 39

foreign companies, 62, 66, 70, 86

foreign direct investment, 18

formation, 90, 129

fouling, 129

foundations, 93

France, 1

freezing, 134

freshwater, 125, 127, 135, 139, 142, 143, 146, 160

frost, 111

fuel cell, 28

fuel prices, 119

funding, 10, 24, 46, 66, 73, 79, 87, 88, 89, 91, 92, 94, 98, 103, 210, 212 funds, 6, 13, 77, 87, 90, 91, 93, 94, 99, 102, 103

fungi, 131

G

gasification, 2, 9, 13, 20, 66, 67, 97

GDP, 2, 12, 13, 14, 17, 106

Generation-based Incentive, 2, 38

German Technical Cooperation (GTZ), 1

Germany, 10, 34, 37, 47, 85, 91, 100, 102, 109, 110

gestation, 51

global climate change, 83

global competition, 102

Global Competitiveness Report, 17, 106

global environment facility, 2

global recession, 40

global warming, 128

Global Wind Energy Council, 2, 35

glycerin, 211

google, 107

governance, 31

Government of India, v, vii, 4, 6, 8, 10, 11, 12, 29, 31, 65, 75, 76, 77, 79, 87, 89, 97, 100, 106, 108, 111, 112, 114, 115, 116, 117, 118, 119, 205, 215

governments, 30, 31, 43, 58, 72, 73, 76, 78, 84, 87, 101, 102, 103, 104, 155

grants, 76, 79, 91, 93, 95, 101, 102, 103, 104, 210, 211

grazers, 131

grazing, 125, 153, 161

green alga, 127

green buildings, 56

greenhouse, 2, 5, 12, 123, 128

gross domestic product, 2, 12

groundwater, 125, 135, 136, 139, 141, 142, 143, 144, 160

growth, vii, 4, 5, 6, 7, 8, 10, 11, 12, 14, 16, 18, 19, 21, 22, 25, 34, 36, 37, 40, 41, 42, 46, 53, 76, 81, 83, 85, 94, 99, 101, 102, 104, 106, 125, 128, 129, 130, 131, 132, 134, 140, 145, 148, 151, 159, 160, 161, 206

growth rate, 16, 19, 21, 34, 37, 53, 85, 101, 128, 131, 151

growth temperature, 151

guidance, 73, 213, 214

guidelines, 28, 31, 37, 44, 89, 117, 119

H

habitat, 6, 57

harvesting, 68, 129, 130, 131, 146

Hawaii, 104, 121

health, 96

heavy metals, 131

hectare, 2, 124, 128, 129, 157

height, 35

hexane, 130

hiring, 214

homes, 94, 95

host, 13, 14

hot spots, 152

hotel, 55

hotels, 55

housing, 74

hub, 7, 35, 39, 45

human, vii, 11, 27, 76, 78, 143, 160, 205, 213

human resource development, 27, 78

human resources, 213

hybrid, 25, 26, 52

hydrocarbons, 203

hydrogen, 28, 203

I

Iceland, 104

ideal, 128

identification, 161

image, 49, 200

images, 109, 110, 140

imports, vii, 12, 21, 39, 46, 124, 126, 212

improvements, 1, 5, 12, 23, 25, 35, 61, 85, 96, 98, 103

impurities, 130

income, 8, 18, 55, 69, 70, 74, 79, 80, 87, 94, 102

income tax, 8, 69, 87, 102

India Renewable Energy Development Agency, 2

Indians, 9

individuals, 65

Indonesia, 23, 102, 104

induction, 129, 212

Industrial Development Bank of India, 2, 94

industrial sectors, 8, 53, 63

industrial wastes, 67

industrialized countries, 13, 86, 97, 99

industries, 9, 10, 25, 55, 63, 64, 67, 68, 70, 72, 75, 78, 81, 82, 83, 84, 148, 207

industry, 4, 8, 9, 19, 25, 26, 34, 35, 40, 41, 55, 71, 81, 82, 83, 105, 108, 110, 124, 125, 126, 128, 130, 131, 142, 153, 157, 161, 208, 209, 211, 212
infancy, 131
information technology, 90
infrastructure, 5, 12, 14, 17, 23, 24, 25, 60, 75, 76, 77, 82, 87, 94, 103, 129, 131, 148, 209, 210, 213
Infrastructure Development Finance Corporation, 2, 96
initiation, 100
institutions, 2, 6, 10, 28, 71, 72, 74, 79, 84, 87, 88, 93, 94, 98, 124, 126, 208, 211, 213
integrated gasification combined cycle, 2, 13
integration, 18, 61, 82
intellectual property, 27, 83
intellectual property rights, 27
Inter-American Development Bank, 103
interest rates, 93, 94
internal rate of return, 44
International Electrotechnical Commission, 2, 45
International Energy Agency, 2, 9, 36, 77, 200
International Finance Corporation, 2, 96, 120
international financial institutions, 95, 99
international relations, 27
International Renewable Energy Conference, 2, 10, 11, 99, 100
international standards, 212
investment, 5, 7, 10, 12, 18, 23, 24, 28, 29, 37, 46, 48, 50, 60, 82, 84, 91, 92, 94, 101, 102, 211
investments, 10, 28, 31, 40, 71, 89, 91, 92, 93, 96, 98, 103
investors, 18, 25, 37, 40, 41, 44, 48, 53, 78, 84, 86, 89, 93, 94, 103, 110, 124, 126
IPR, 119
Ireland, 104
iron, 81, 127, 148
irradiation, 4, 45, 53
irrigation, 71, 136, 139, 143, 144, 146, 148, 157
Israel, 53, 54, 55
issues, 4, 28, 84, 89, 131, 142
Italy, 102, 104

J

Japan, 28, 53, 54, 102, 103
Jawaharlal Nehru National Solar Mission, 2, 7, 12, 107, 108, 110, 111, 113, 118, 120
job creation, vii, 124, 126
joint ventures, 210
Jordan, 102

jurisdiction, 31, 45, 155

K

kerosene, 75, 80, 96
Kreditanstalt für Wiederaufbau (German Development Bank), 2

L

lakes, 147, 148
laminar, 128
land acquisition, 155
landscape, 11, 125, 161, 203
laws, 18, 28
lead, 37, 62, 66, 83, 85, 103, 148, 206
leadership, 5, 11, 12, 29, 101
lending, 6, 28, 92, 96, 97, 210
liberalization, 18, 31, 88
life cycle, 128, 129, 131
light, 81, 128, 129, 156
lipids, 126, 127, 131
liquefied petroleum gas, 2, 8, 55
liquids, 146, 203
livestock, 143, 153, 158
living conditions, 76
loan guarantees, 99, 103
loans, 9, 23, 63, 73, 91, 92, 93, 94, 96, 97, 98, 103, 210
local conditions, 66
local government, 101
localization, 85, 86
logging, 68
low temperatures, 52
low-interest loans, 8, 62, 77, 95, 102
LPG, 2, 55, 76

M

machinery, 88, 128, 210
macroalgae, 127
magnitude, 5, 130, 132
majority, 6, 10, 12, 14, 17, 18, 30, 55, 63, 71, 75, 91, 148, 151, 155
management, 14, 39, 49, 61, 79, 131, 140
manpower, 213, 214
manufacturing, 10, 19, 40, 41, 42, 45, 46, 51, 52, 66, 84, 85, 86, 89, 105, 148, 212
manufacturing companies, 46

manure, 9, 63, 64, 65, 151, 158
market share, 34, 40, 98
marketing, 25, 209
marketplace, 98
mass, 4, 41, 67, 127, 131, 134, 206
materials, 11, 63, 129, 207, 211
measurement, 31
meat, 63, 64
media, 110, 112, 120, 121, 145
melting, 137
mentoring, 94
messages, 100
metals, 103
meter, 31, 124
methodology, 161
Mexico, 104
MFI, 2, 95
microfinance institutions, 2, 94
Middle East, 2, 52, 104
Middle East and North Africa, 2, 104
migration, 44, 111
milligrams, 143
Ministry of New and Renewable Energy, 2, 6, 24, 27, 123, 200, 211, 214, 215
Ministry of Petroleum and Natural Gas, 3, 73, 123, 199, 200
Ministry of Power, 3, 6, 28
Ministry of Rural Development, 3, 72, 76, 123, 153, 200, 211, 213, 214
Minneapolis, 117
mission, 27, 31, 43, 73, 105, 107, 108, 110, 111, 113, 118, 119, 120
missions, 6
mixing, 128, 129
models, 36, 78, 79, 130
modernization, 3, 14, 18, 23
modifications, 71
modules, 45, 46, 47, 120
molasses, 9, 27, 63, 71, 72, 207, 208
molecules, 132
momentum, 100
MSW, 3, 14, 66
multinational companies, 84
municipal solid waste, 3, 14

N

National Action Plan on Climate Change, 3, 5, 12, 13
National Aeronautics and Space Administration, 3, 43

National Bank for Agriculture and Rural Development, 3, 63, 78
National Biogas and Manure Management Program, 3, 65
National Electricity Policy, 3, 6, 21, 29, 76, 108
national policy, 29
National Renewable Energy Laboratory, v, 1, 3, 43, 123, 215
National Thermal Power Corporation, 3, 21, 43, 107
natural disaster, 135
natural disasters, 135
natural gas, 26, 66, 86, 104, 125, 142, 146, 160
natural resources, 4, 96
Nepal, 96
Netherlands, 119, 200
New Zealand, 104
next generation, 206, 207
NGOs, 59, 78, 79, 92, 94
nitrogen, 127, 132, 151, 160
non-governmental organization, 3, 59
North Africa, 2, 52, 104
NTPC Vidyut Vyapar Nigam Ltd, 3, 43
nutrient, 128, 131, 147, 151
nutrients, 125, 127, 128, 132, 147, 148, 151, 157, 159, 160, 161

O

oil, vii, 3, 4, 5, 9, 11, 12, 14, 27, 63, 70, 71, 72, 83, 100, 104, 124, 125, 126, 127, 128, 129, 130, 131, 141, 142, 146, 150, 157, 160, 161, 203, 205, 207, 208, 209, 210, 211
oilseed, vii, 72, 124, 126, 211, 213
omega-3, 131
open spaces, 82
operating costs, 59, 135
operations, 9, 10, 41, 63, 65, 78, 80, 97, 130
opportunities, vii, 4, 7, 10, 11, 13, 14, 25, 32, 40, 41, 52, 53, 62, 64, 70, 75, 81, 84, 85, 100, 104, 124, 126, 151, 157, 158, 160, 161, 210, 213
optimal performance, 29
optimization, 131
organic compounds, 146
outreach, 53, 62, 74
overlap, 141, 149, 151
oversight, 24
ownership, 87, 148, 155
oxygen, 130

P

Pacific, 1, 90, 108, 119
parity, 11, 44, 45, 85, 105
participants, 91, 100
pasture, 125, 151, 153, 161
patents, 90, 211
pathogens, 131
pathways, 130
payback period, 55
peat, 106
penalties, 29, 38
percolation, 129
permeability, 151
petroleum, vii, 2, 8, 9, 27, 55, 71, 117, 124, 126, 127, 128, 130, 203
Petroleum, 3, 60, 73, 117, 123, 199, 200, 213, 214
petroleum-based fuels, vii, 124, 126, 128
pH, 127
Philippines, 102, 104
phosphates, 147
phosphorous, 127, 132, 151, 160
phosphorus, 151
photosynthesis, 128
physiology, 127
plants, 7, 8, 9, 10, 11, 13, 17, 20, 23, 25, 33, 37, 40, 41, 45, 47, 49, 50, 51, 52, 53, 57, 58, 60, 62, 63, 64, 65, 66, 67, 68, 70, 76, 81, 82, 86, 87, 94, 103, 121, 127, 132, 148, 151, 157, 158, 160, 210, 212, 213
platform, 81, 101, 102
playing, 60, 75, 84, 99, 207
Poland, 102
policy, vii, 4, 6, 7, 11, 12, 14, 24, 28, 29, 30, 31, 36, 45, 51, 52, 73, 91, 100, 101, 102, 108, 117, 200, 213
policy initiative, 51
policymakers, 104, 124, 125, 126, 153, 161
pollution, 147, 206
ponds, 124, 128, 129, 135, 145, 146, 151, 156, 157, 161
population, vii, 4, 9, 11, 13, 14, 16, 18, 64, 74, 75, 124, 126, 140, 148, 160
porosity, 151
portfolio, 98, 102
poultry, 65, 70, 143
poverty, 1, 7, 9, 29, 30, 75
poverty line, 1, 7, 9, 29, 30, 75
Power Finance Corporation, 3, 28

power generation, 9, 11, 13, 17, 19, 20, 21, 22, 23, 28, 29, 31, 39, 47, 50, 51, 63, 66, 67, 68, 69, 80, 81, 82, 83, 86, 87, 91, 92, 93, 94, 96, 109, 126, 131, 148
power plants, 10, 12, 17, 18, 19, 21, 23, 25, 37, 40, 46, 47, 51, 52, 60, 63, 67, 75, 81, 82, 83, 85, 102, 103, 135, 148, 151, 157, 160
Power Purchase Agreement, 3, 44
practice guides, 89
precipitation, 132, 135, 137
predators, 131
President, 100
prevention, 129, 147
private banks, 95
private enterprises, 84
private investment, 18, 28
producers, 18, 28, 30, 37, 38, 63, 70, 71, 74, 81, 93, 117
product design, 86
product market, 131
product performance, 212
production costs, 39, 86, 130
productivity rates, 128
professionals, 89, 104
profit, 37
profitability, 89
project, 8, 10, 13, 37, 38, 39, 40, 44, 45, 46, 48, 49, 50, 51, 52, 56, 59, 60, 62, 65, 66, 67, 69, 70, 74, 79, 80, 84, 85, 89, 90, 91, 93, 94, 97, 98, 99, 101, 102, 104, 118, 126, 155, 159
promoter, 51
property rights, 84
protected areas, 152, 159
protection, 83, 90, 96
public awareness, 74
public financing, 100
public interest, 27
public investment, 101
public sector, 17, 28, 66, 83, 84, 103
pulp, 55, 63, 64, 67, 68, 148
pumps, 19, 43, 76, 88
purchasing power, 88
purification, 66, 67, 129, 130

Q

quotas, 101, 102

Index 225

R

radiation, 8, 41, 132, 133, 156, 159, 160, 162
Radiation, 43, 49
rainfall, 135, 137, 139, 148
ramp, 51
rate of return, 13
reality, 129
recommendations, iv, 60, 123
recovery, 66
recycling, 46, 128, 160
reform, 24, 31
Reform, 23
Registry, 209
regulations, 7, 29, 31, 32
regulatory requirements, 131
relevance, 10
reliability, 56, 61, 86, 212
remote sensing, 49
Remote Village Electrification, 3, 78
renewable energy, vii, 3, 4, 5, 6, 7, 9, 10, 11, 12, 14, 17, 19, 21, 24, 25, 26, 27, 28, 29, 30, 31, 32, 33, 34, 37, 38, 40, 43, 60, 61, 74, 75, 76, 77, 78, 81, 83, 84, 86, 87, 88, 89, 90, 91, 92, 93, 94, 95, 96, 97, 98, 99, 100, 101, 102, 103, 104, 107, 113, 118, 205
Renewable Energy and Energy Efficiency Partnership, 3, 96
Renewable Energy Certificates, 3
Renewable Energy Policy Network for the 21st Century, 1, 3
renovation and modernization, 3, 23
repair, 23, 40
requirements, 7, 14, 17, 18, 29, 32, 66, 68, 75, 82, 85, 86, 125, 127, 148, 151, 159, 160, 161, 206, 209, 212, 213
research and development, 3, 9, 10, 27, 41, 47, 49, 123
research institutions, 74, 212, 213
reserves, 111, 148
residues, 25, 27, 63, 67, 68, 71, 131, 207
resolution, 42, 43, 49
resource availability, 60, 78
resources, vii, 4, 5, 11, 12, 13, 18, 27, 28, 29, 33, 39, 49, 63, 67, 68, 69, 72, 75, 76, 81, 90, 101, 102, 106, 125, 126, 127, 128, 131, 132, 135, 139, 140, 141, 142, 144, 146, 150, 160, 161, 164, 205, 206, 207, 210
response, 40, 41, 127
restrictions, 81

restructuring, 15
retail, 209
revaluation, 111
revenue, 76, 78, 80, 99, 131
rice husk, 68
risk, 86, 93, 102, 129
risks, 84, 87, 89, 91, 94, 99
river basins, 137
river systems, 136
Romania, 102
roots, 127
royalty, 88
runoff, 147, 161
rural areas, 5, 9, 17, 18, 19, 28, 55, 63, 64, 65, 66, 75, 76, 78, 90, 148
rural development, vii, 124, 126, 206
Rural Electricity Distribution Backbone, 3
rural population, 9, 206
Russia, 136

S

saline water, 127, 135, 143, 145
salinity, 127, 134, 143, 146, 148
salts, 145, 146
Saudi Arabia, 201
savings, 53, 65
scaling, 11
school, 70
science, 89, 100
scope, 67, 86, 94, 159, 207
scripts, 110
sea-level, 134
second generation, 71, 207, 210, 211
security, vii, 4, 5, 11, 27, 41, 70, 75, 104, 124, 126, 205, 206
sediment, 61
seed, 208
seedlings, 208
sellers, 18
seminars, 90
service provider, 30, 60
services, 5, 13, 28, 62, 75, 85, 86, 92, 94, 148, 214
sewage, 64, 66, 123, 146, 148, 157, 158
short supply, 72
shortage, 4, 13, 73, 81, 137
shortfall, 209
showing, 41, 49
shrubs, 207, 208
silicon, 45, 46, 85, 151

Singapore, 91
sludge, 151
small businesses, 70
social change, 14
social responsibility, 2
society, 81, 84
soil type, 151
Solar and Wind Energy Resource Assessment, 3, 43
solar cells, 46
solar collectors, 52
Solar Energy Centre, 3, 6, 28
solution, 125, 146, 157, 161
solvents, 130
South Africa, 23
South America, 39, 103
South Asia, 120
South Korea, 54, 102, 104, 121
Spain, 34, 37, 47, 52, 102, 109, 113, 121
special incentive package scheme, 3
specialists, 60
species, 127, 130, 131, 134, 144, 146, 148, 208
specifications, 130
spending, 102
stability, 49, 129, 131
stabilization, 124, 157
stakeholders, 11, 74, 75, 99, 100, 208, 212
starch, 63, 65, 207
state, 8, 9, 18, 23, 31, 32, 33, 34, 37, 38, 40, 42, 43, 45, 47, 51, 52, 58, 60, 61, 68, 69, 70, 72, 73, 74, 76, 77, 78, 82, 87, 91, 93, 101, 102, 104, 116, 118, 124, 125, 126, 135, 153, 155, 157, 161, 211
State Electricity Board, 3, 18
State Electricity Regulatory Commission, 3, 6, 28
states, 7, 8, 9, 12, 17, 21, 24, 28, 31, 32, 33, 34, 37, 38, 45, 49, 55, 58, 59, 65, 67, 68, 69, 73, 76, 82, 85, 87, 97, 102, 104, 125, 132, 133, 134, 135, 139, 151, 155, 161
statistics, 81, 106
steel, 81, 129, 148, 157, 158, 160
steel industry, 157
stimulus, 103
stock exchange, 91
storage, 47, 49, 52, 57, 82, 140, 209, 210
storage media, 47
stoves, 78, 79
strain improvement, 131
stress, 127
structural changes, 4
structure, 31
subsidy, 7, 59, 77, 78

substitutes, 27, 126, 130
sugar beet, 71, 207, 211
sugar industry, 9, 27, 67, 70, 71, 81, 86, 207, 208
sugar mills, 67, 68, 73
sugarcane, 68, 71, 208, 209, 211
sulfur, 151
Sun, 50, 121
supplementation, 151
suppliers, 46, 73, 79, 97
supply chain, 49, 55
supply disruption, 11
surface area, 156
Surface Meteorology and Solar Energy, 3, 43
surplus, 9, 47, 63, 67, 68, 70, 71, 81, 209
sustainability, 9
sustainable development, 92, 97, 206
sustainable energy, 75
Sustainable Energy Finance Initiative, 3, 91
Sweden, 102
Switzerland, 106
synthesis, 203

T

takeover, 155
tangible benefits, 78
target, 7, 8, 25, 41, 43, 51, 56, 58, 72, 73, 101, 102, 108, 144, 206
tariff, 6, 7, 8, 9, 28, 29, 33, 38, 43, 44, 45, 47, 49, 51, 52, 60, 69, 121
tax deduction, 103
tax incentive, 73, 90
taxes, 44, 73, 87, 210, 212
taxis, 74
techniques, 129, 159
technological developments, 39
technology, 4, 5, 6, 7, 8, 9, 10, 11, 12, 13, 21, 23, 26, 27, 29, 35, 37, 39, 41, 42, 45, 47, 51, 52, 53, 55, 62, 66, 67, 70, 78, 79, 83, 84, 85, 86, 87, 88, 89, 90, 91, 92, 93, 94, 98, 99, 102, 103, 104, 112, 124, 125, 126, 127, 128, 130, 131, 141, 142, 144, 146, 152, 155, 159, 160, 161, 211, 212
technology transfer, 45, 83, 84, 87, 88, 90
temperature, 52, 56, 83, 127, 128, 129, 131, 132, 134, 135, 203
territory, 84
The Energy and Resources Institute, 3, 23
theft, 23
thermal energy, 69, 83, 104
time periods, 47

Index

total costs, 103
total energy, 62
total product, 71
trade, 13, 28, 33, 119
training, 10, 28, 66, 70, 74, 78, 84, 89, 213
trajectory, 4, 18
transaction costs, 99
transactions, 91
transesterification, 130, 209
transformation, 98
transmission, 6, 18, 24, 28, 29, 30, 31, 52, 70, 93, 96, 104, 108
transport, 13, 104, 126, 206, 207, 210
transportation, 4, 6, 9, 26, 27, 71, 74, 98, 125, 126, 129, 130, 131, 136, 148, 150, 151, 155, 157, 159, 161, 205, 206, 211, 213
transportation infrastructure, 130, 155, 213
treatment, 64, 66, 123, 125, 131, 144, 146, 147, 148, 151, 157, 158, 161
Turkey, 53, 54, 104
turnover, 62

U

U.S. Geological Survey, 201
U.S. National Aeronautics and Space Administration, 3, 43
United, 1, 3, 10, 34, 37, 39, 52, 54, 87, 90, 91, 95, 97, 100, 101, 102, 103, 104, 109, 116, 117, 118, 135, 136, 148, 199, 201
United Kingdom (UK), 70, 87, 102, 118, 119, 200
United Nations (UN), 3, 87, 90, 91, 95, 97, 100, 117, 118, 148, 201
United Nations Development Programme, 3
United Nations Environment Programme, 3, 91, 95, 97
United Nations Framework Convention on Climate Change, 3
United States, 1, 3, 10, 34, 37, 39, 52, 54, 91, 101, 102, 103, 104, 109, 116, 117, 135, 136, 199
United States Agency for International Development, 3
universities, 88, 93, 213
urban, 5, 9, 20, 27, 28, 55, 64, 66, 148
urban areas, 55, 148
urban population, 64
Uruguay, 102
USDA, 117

V

value added tax, 101
variations, 132, 134, 135, 141
varieties, 211
vegetable oil, 126, 207
vehicles, 26, 74, 88, 94
venture capital, 10, 91, 94
viruses, 131
volatility, 5

W

waiver, 94
war, 107
Washington, 4, 100, 101, 104, 119, 201
Washington International Renewable Energy Conference, 4, 100
waste, 9, 11, 20, 25, 27, 28, 63, 64, 65, 66, 67, 70, 72, 92, 98, 124, 126, 128, 135, 157, 207
water, vii, 3, 6, 7, 8, 19, 38, 41, 42, 43, 46, 53, 55, 57, 62, 71, 76, 82, 83, 90, 113, 124, 125, 126, 127, 128, 130, 131, 132, 135, 136, 137, 139, 141, 142, 143, 144, 145, 146, 148, 151, 155, 157, 158, 159, 160, 161, 199, 201
waterways, 4
web, 118
well-being, 12
wells, 142, 146
wetlands, 159
wind farm, 4, 82, 94
wind speeds, 11
wind turbines, 34, 36, 39, 40, 81, 88, 89
Wisconsin, 200
wood, 8, 17, 62, 65, 68, 75, 106, 207
wood waste, 207
workers, 65
World Bank, 28, 87, 89, 90, 95, 96, 98, 103, 107
worldwide, 49, 53, 91, 100, 127, 128

Y

yeast, 65
yield, 9, 62, 71, 78, 125, 128, 150, 157, 161